Computational Intelligence in Robotics and Automation

This book will help readers to understand the concepts of computational intelligence in automation industries, industrial IoT (IIOT), cognitive systems, data science and E-commerce real-time applications.

The book:

- covers computational intelligence in automation industries, IIOT, cognitive systems and medical imaging;
- discusses intelligent robotics applications with the integration of automation and artificial intelligence;
- covers foundations of the mathematical concepts applied in robotics and industry automation applications;
- provides application of artificial intelligence (AI) in the area of computational intelligence.

The text covers important topics including computational intelligence mathematical modeling, cognitive manufacturing in industry 4.0, artificial intelligence algorithms in robot development, collaborative robots and IIoT, medical imaging and multi-robot systems.

Discussing the advantages of the integrated platform for industry automation, robotics and computational intelligence, this book will be useful for graduate students, professional and academic researchers in the fields of electrical engineering, electronics and communication engineering, and computer science. It enlightens the foundations of the mathematical concepts applied in robotics and industry automation applications.

Computational Intelligence in Robotics and Automation

Edited by
S. S. Nandhini
M. Karthiga
S. B. Goyal

CRC Press
Taylor & Francis Group
Boca Raton London New York

CRC Press is an imprint of the
Taylor & Francis Group, an **informa** business

First edition published 2023
by CRC Press
6000 Broken Sound Parkway NW, Suite 300, Boca Raton, FL 33487-2742

and by CRC Press
4 Park Square, Milton Park, Abingdon, Oxon, OX14 4RN

CRC Press is an imprint of Taylor & Francis Group, LLC

ISBN: 978-0-367-75449-5 (hbk)
ISBN: 978-1-032-02061-7 (pbk)
ISBN: 978-1-003-18166-8 (ebk)

DOI: 10.1201/9781003181668

Typeset in Times
by SPi Technologies India Pvt Ltd (Straive)

Contents

Preface

The areas of computational intelligence, robotics and automation have their own never-ending applications. This is an interesting and useful technology everyone should know. In day-to-day life, knowingly or unknowingly, we are using many automated applications and the same has to be explored. Also, many common applications are available where automation is implemented. The main aim is to explore all the intelligence behind the applications of automation deploying robots and also a wide range of upcoming automation in industries.

Many of us are already using automated products/processes in our day-to-day life, but speaking about automation and its importance in a technological world, we are not aware of the daily products that we are using, personally deploy robots and achieve automation. One such example is e-bots in customer services. Most of the websites deploy a chatbot to guide the users and to give clarifications on their queries. These are the software bots available in many online customer services that we are using unknowingly. Whenever we think of automation and robotics, we have an illusion that they are used in high-end technologies and large-scale industries. But the fact is that automation is in our common areas like agriculture, medicine, healthcare, insurance, bank-sectors and so on. We should explore such areas where automation is already being implemented and potential areas where we can introduce automation to ease the task and to make it more efficient.

Apart from this, the book gives information about the technologies behind automation by using case studies in the semi-automated industries like automotive industries where physical robots are used to automate the assembling of automobiles, painting them, removing and adding machinery parts, etc. So, these are some of the main reasons to write about robotics and automation and the intelligence behind them.

This book aims at giving the technological description behind computational intelligence, automation and robotics in various fields of applications. Also, it will make the readers have a clear idea about robotics and automation. Further, it will make them understand whether robotics and automation are different and also answer the question of whether one can exist without the other.

Software Robotics Process Automation is widely used in customer services to handle many common queries. In case, a customer has a query that is categorized as very common by the organization and the same will be handled by robots mostly tagged as e-bots. Only if they can't handle the query from the customer, it will get forwarded to human responders in customer services. This makes the customer not wait so long to get the response and also assures 100% quality responses. This is a prevalent application; almost every one of us might have experienced it.

The physical robots are mostly used in automotive industries. Because, the work in automotive industries are mostly routine and hence replacing human with robots is easier. The most widely known tasks for robots in the automotive industry include painting, welding, assembling automobiles, transporting the parts within the industry, etc.

This book is a useful resource to attract works on multidisciplinary spanning across the electronics engineering, electrical engineering, computer science and engineering, medical and health science, environmental studies, industrial engineering on technologies, novel idea/approaches and visionary ideas related to the domain of computational intelligence, robotics and automation for researchers and industry practitioners; IT managers in computational intelligence, robotics and automation; detection of dental age, personalized care to diagnosis, weather forecasting, AI algorithms for robot development, utilization of RPA in a pandemic situation, IoT-based smart farming/irrigation, haze removal techniques, cyber-machine learning for cyber-attack prediction, applications in automobile industries using mobile robots etc.

Overall, readers will have a better understanding of the technology and after-effects of automation and robotics.

The editors are thankful to the writers for presenting the different techniques, concepts and case studies for computational intelligence, robotics and automation throughout this book. We hope that the book will make a significant contribution to the community by bringing together research on robotics automation and process development. We really hope that many more people will join us in this vital effort. Good luck with your reading!

S. S. Nandhini
BIT, Tamil Nadu, India

M. Karthiga
BIT, Tamil Nadu, India

S. B. Goyal
City University, Malaysia

Acknowledgments

We are grateful to the chapter contributors and reviewers for shaping the chapters of this book. Further, we would like to thank our employer, colleagues and students for rendering their support toward editing this book.

Editors

S. S. Nandhini is currently working as senior assistant professor and Head of Information Science and Engineering in Bannari Amman Institute of Technology, Tamil Nadu. She has completed her M.E. in Computer Science Engineering from PSG College of Technology, Coimbatore, Tamil Nadu. She is pursuing her PhD from Anna University in Information and Communication Engineering and is presently working in utility mining. She has teaching experience of more than 8 years. She is a distinguished speaker in ACM.

M. Karthiga is currently working as an assistant professor in the Department of Computer science and Engineering, Bannari Amman Institute of Technology, Sathyamangalam. She completed her M.E. in Computer Science Engineering from Kongu Engineering College. She is pursuing her PhD from Anna University in Information and Communication Engineering and is presently working in machine learning and deep learning projects. She is involved in Data Science Lab. She has teaching experience of more than 7 years.

Dr. S. B. Goyal is Director in Faculty of Information Technology, City University, Malaysia. He has M.Sc., M.Tech. and PhD from Banasthali Vidyapith, India. He has teaching experience of more than 20 years. He is IEEE Senior Member. His research areas are Modeling and Simulation, Data Science and Artificial Intelligence.

Contributors

S. Ashwini
Kongu Engineering College
Perundurai, India

B. Aishvarya
Kongu Engineering College
Perundurai, India

Bharath Chandan Reddy
Sreenidhi Institute of Science &
   Technology
Hyderabad, India

E. Balamurugan
University of Africa
Toru-orua, Nigeria

Abhishek Choubey
Sreenidhi Institute of Science &
   Technology
Hyderabad, India

P. Dhivya
Bannari Amman Institute of Technology
Erode, India

Ms. P. Divya
Bannari Amman Institute of Technology
Erode, India

S. B. Goyal
City University
Selangor, Malaysia

P. Harinitha
Kongu Engineering College
Perundurai, India

B. Hemalatha
KGiSL Institute of Technology
Coimbatore, India

C. S. Kanimozhiselvi
Kongu Engineering College
Perundurai, India

K. S. Kannan
CMR Engineering College
Hyderabad, India

M. Karthiga
Bannari Amman Institute of Technology
Erode, India

Hyder Ali Segu Mohamed
College of Engineering and Information
   Technology
Buraydah, Kingdom of Saudi Arabia

Senthil Kumar Muthusamy
University of Technology and Applied
   Sciences
Sultanate of Oman
Muscat, Oman

S. S. Nandhini
Bannari Amman Institute of Technology
Erode, India

P. Naveena
Bannari Amman Institute of Technology
Erode, India

S. K. Nivetha
Kongu Engineering College
Perundurai, India

A. Padmashree
Bannari Amman Institute of Technology
Erode, India

D. Palanivel Rajan
CMR Engineering College
Hyderabad, India

Balachandra Pattanaik
Wollega University
Nekemte, Ethiopia

D. Prabha Devi
Bannari Amman Institute of Technology
Erode, India

T. Pradeepika
Bannari Amman Institute of Technology
Erode, India

V. Praveen
Bannari Amman Institute of Technology
Erode, India

V. Priyadharshini
Bannari Amman Institute of Technology
Erode, India

N. Rajkumar
KGiSL Institute of Technology
Coimbatore, India

P. Rajesh Kanna
Bannari Amman Institute of Technology
Erode, India

R. Ramya
Bannari Amman Institute of Technology
Erode, India

K. P. Sampoornam
Bannari Amman Institute of Technology
Erode, India

V. Santhi
PSG College of Technology
Coimbatore, India

A. Saran Kumar
Bannari Amman Institute of Technology
Erode, India

S. Saranya
Bannari Amman Institute of Technology
Erode, India

G. Sathish Kumar
Sri Krishna College of Engineering and
    Technology
Coimbatore, India

P. Sathish Kumar
Bannari Amman Institute of Technology
Erode, India

N. Senthilkumaran
Vellalar College for Women
Erode, India

Monika Sharma
Galgotias University
Greater Noida, India

R. C. Suganthe
Kongu Engineering College
Perundurai, India

R. S. Soundariya
Bannari Amman Institute of Technology
Erode, India

R. M. Tharsanee
Bannari Amman Institute of Technology
Erode, India

Dileep Kumar Yadav
Galgotias University
Greater Noida, India

D. Yuvaraj
Bannari Amman Institute of Technology
Erode, India

1 Automatic Detection of Dental Age Assessment Using an Efficient Elman Neural Network with Dragonfly Optimization

B. Hemalatha and N. Rajkumar

KGiSL Institute of Technology, Coimbatore, India

CONTENTS

DOI: 10.1201/9781003181668-1

1.1 INTRODUCTION

In recent times, individuals' migration leads to rise in necessity of forensic age estimation, specifically where birth dates are not documented. Age estimation is essential for asylum regulations, as there exists a diverse judicial process based on individuals' age. Moreover, forensic age estimation is an essential factor in penal or civil law procedures. Till date, the suggested form of DA estimation for forensic computation in adolescents includes external body examination, skeletal stage assessment and dental development. DA estimation with relevant age includes thresholds. In European countries 13–21 years are considered the age for third molar development, which is done with panoramic X-ray, that is, an Orthopantomogram (OPG). Numerous stagings have been modeled with diverse morphological criteria of tooth development. An extensively known method is sourced on mineralization assessment and third molar eruption.

Every stage reference values for diverse populations as per stages are modified to age estimation with years. Reference values and staging systems are considered for optimization and researches; the baseline for dental development evaluation is turned out to be a major research issue. The critical reason is that an OPG depends on ionizing radiation that is legitimately restricted in numerous realms for non-clinical signs. With general medical problems, ionizing radiation utilization has been disputable. MRI is not related to X-rays and also with ionizing radiation exposure. It gives focal point evaluation of pictures with superposition and empowering 3D post preparation. Till time, MRI is not exploited in dentistry even if it is considered in initial studies, for example, to identify dental and jaw abnormalities and determined to be appropriate for adolescents and children.

Unknown individuals' age can be measured by physical correlation, dental maturity and skeletal maturity. Numerous radiological approaches include dental maturity as an indicator that is extensively studied. This comprises age estimations sourced on open teeth apices measurement, pulp tooth ratio and stages of tooth development. Demirjian and his investigators portrayed Tooth Development Stages (TDS) as utmost easier and consistent approach due to uppermost standards for both interviewer and intra-viewer arrangement.

Numerous dental age (DA) estimation approaches have been investigated by researchers. Demirjian's method is extensively utilized and approved for DA estimations based on eight development stages of tooth acquired from radiographic images for forensic age estimation purpose. As Demirjian scores is used to individuals with diverse population samples to examine reliability and applicability of this approach and reported as overestimation of age in native population samples. Therefore, investigators, especially in Willems et al., anticipated modified dental scores to perform DA estimation in simpler and in accurate manner. In legal preparation, Willems approach has been a steady methodology for DA assessment and various experts have incorporated flexibility and materialness of dental scoring approach in vast populace.

However, researchers said that age detection and classification accuracy are extremely complex in the above-mentioned method. Therefore, to resolve this crisis, image processing techniques have been applied for assessing the age of a person automatically. In this, Elman Neural Network with Dragonfly Optimization ENN–DO-based classification scheme is anticipated and performance outcomes are contrasted with the prevailing age-prediction methods.

1.2 BACKGROUND

1.2.1 DENTAL AGE ASSESSMENT

Age assessment is valuable for public, criminal, legal and civilization purposes. Numerous strategies have been conceived to assess sequential age. These incorporate physical development estimations and others that depend on dental turn of events. Tooth advancement for age assessment has been utilized for quite a while. The physical advancement of every individual is influenced by hereditary, wholesome, climatic, hormonal and natural components, yet considerably under the impacts of extraordinary fundamental sickness, dental improvement seems, by all accounts, to be influenced uniquely to a minor degree. Tooth advancement shows less changeability than other formative highlights and displays low fluctuation corresponding to the ordered age. Age can be assessed in kids and teenagers by the improvement of deciduous and perpetual teeth, including third molar, as long as 26 years. After this, age must be evaluated by underlying changes in teeth.

Scientific age assessment is a part of legal science continually advancing because of the rising quantities of asylum seekers lacking substantial individuality papers. This issue is especially important on account of youths and youthful grown-ups engaged with common and criminal methodology or looking for refuge. In these cases, it is vital to utilize age-assessment strategies which permit the operators to decide as precisely as conceivable the legitimately applicable ages which change as indicated by the public. There are three measures used for assessing the age:

1. Actual assessment is done through recognition of anthropometric estimations (mass, stature, constitution), indications of erotic development examination, ID of improvement infections;
2. Bone age assessment for the subject is not shown up effectively since at the age of 17 the bones get fused and also through the clavicle bone, assessment of age is not evaluated due to the hardening part of epiphysis bone. Based on the above-given fact, these are not explicitly suggested.
3. Dental assessment can be done by dentist based on the tooth development and testing the condition of dentition using X-ray.

Numerous approaches of DA assessment were planned by many investigators for developing folks. In the above-mentioned approaches, radiographs were used as evidence to analyze the progressive sequence of teeth development, and each stage of development was coded and scored. These scores were manipulated meticulously

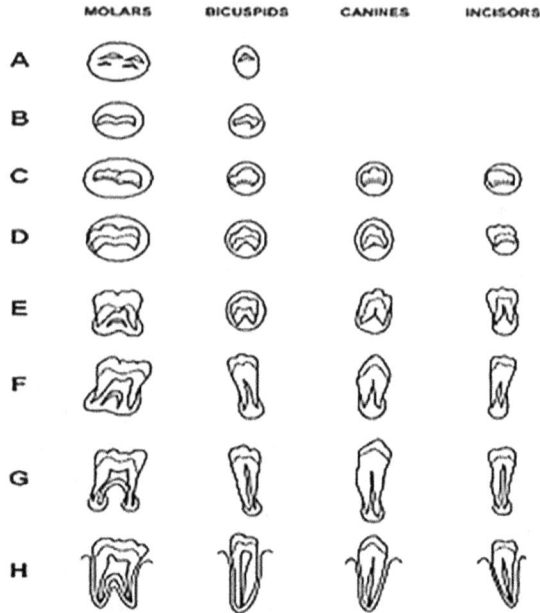

FIGURE 1.1   Tooth development stages (Demirjian method).

to derive the DA of an individual and compared by their CA, with acceptable error limits. However, all these methods dated back a few decades and the change in the growth trend of the current generation alarmed for formulating a newer method of DA assessment. Figure 1.1 depicts TDS that are modeled by Demirjian.

1.2.2   Antiquity of Age Assessment

The primary endeavor was prepared in England to utilize teeth as a marker for assessing the age. Cutting-edge of the mid-19th century, in light of financial downturn, adolescent work and culpability were not kidding social issues.

1.2.2.1   Necessity for Age Assessment

Each individual whether in any condition has option to be recognized for a few reasons:

Living individual requires age assessment
Dead person requires age estimation

1.2.3 METHODS OF AGE ESTIMATION IN ADULTS

1.2.3.1 Histological Methods

1.2.3.1.1 Dentinal Translucency

At first, investigators utilized dentinal clarity strategy only for age assessment. Investigators presumed that clarity of the teeth root increments by period to get noticeable during the third decade of life starting on peak then progressing coronally. It remains because of diminishing breadth of dentinal tubules brought about by expanded intratubular calcification.

1.2.3.1.2 Incremental Lines of Cementum

Assess the age by utilizing a cellular cementum gradual appearances. It is additionally recommended that hypomineralized groups in these steady lines give a sign of circumstances, such as gestations, bone injury and renal turmoil. Significant weakness of this technique is that it expects teeth to be removed. These incremental lines of cementum are more useful for finding the age of dead individuals.

1.2.3.1.3 Typical Age of Erosion

Based on the incremental lines of cementum, the investigators showed the age with level of wearing down and with age level of weakening.

1.2.3.1.4 Newborn Line in Enamel and Dentin

Neonatal line is measured as a natal pointer. Age can be assessed by estimating the thickness of enamel and dentin from the newborn line and then partitioned via every day pace of development.

1.2.3.1.5 Racemization Method

They assessed the time of living individual utilizing racemization strategy in dentinal biopsy cases. The strategy arose as an in vivo procedure.

1.2.3.2 Biochemical Methods

1.2.3.2.1 Amino Acid Racemization

Amino acids are the body building blocks which help in protein binding. Aspartic corrosive can be realized amino corrosive has quick pace of racemization for example it gets precipitously changed over starting with one sort then onto the next with expanding age. They are primarily set up in brain, crystalline lens bone and teeth. Nonetheless, it is high in root dentine. Accordingly, fangs remain viewed as significant wellspring old enough assessment in grown-ups.

1.2.3.3 Radiological Method

1.2.3.3.1 Schour and Masseler Method

The method has contemplated the advancement of deciduous and perpetual teeth, depicting ordered strides from 4 months to 21 years old utilizing radiographs and distributing the mathematical improvement outlines for individuals.

1.2.3.3.2 *Demirjian, Goldstein and Tanner Method*

Demirjian, Goldstein and Tanner appraised seven mandibular perpetual teeth in the request for second molar (M2), first molar (M1), second premolar (PM2), first premolar (PM1), canine (C), lateral incisors (I2) and central incisor (I1) and decided eight phases (A to H) of tooth mineralization along with stage zero. The development score apportioned to the teeth is added and a complete development score is acquired. This complete is then subbed in a formula. Remembering the distinctions in the dental improvement of females and males, separate development scores and equations were accommodated for males and females.

1.3 PROPOSED METHODOLOGY

This section discusses in detail about anticipated ENN–DO-based classification and estimation. Here, pre-processing, effectual segmentation, feature extraction and classification have been carried out for improving precision of age identification. The anticipated model is well organized and evaluated in guaranteeing sub-segments. Figure 1.2 depicts the prediction strategy in detail.

1.3.1 SYSTEM OVERVIEW

The model presented here comprises four phases such as image acquisition, preprocessing, segmentation, feature extraction and classification. The entire stage by stage anticipated model is presented in Figure 1.2.

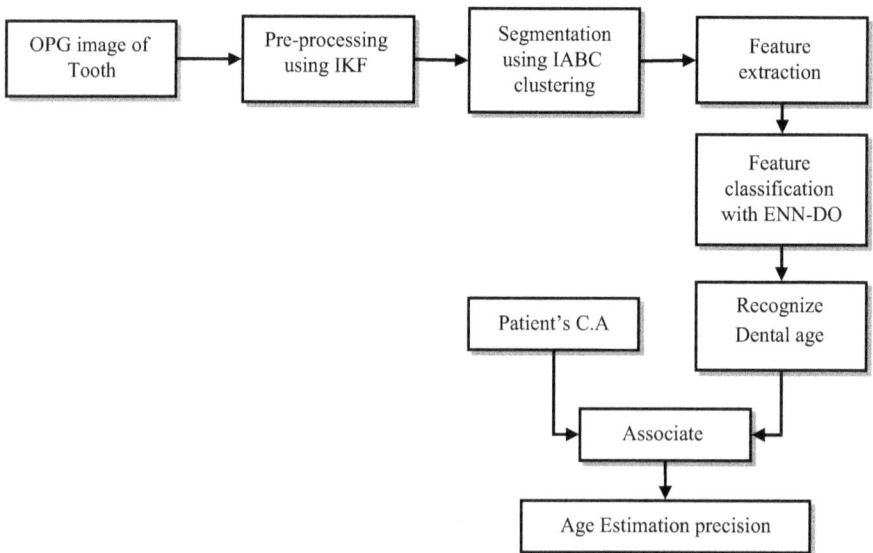

FIGURE 1.2 Design of anticipated ENN–DO method.

Image acquisition: The original OPG teeth image is accumulated from Kovai Scan Center, Coimbatore.

Pre-processing: Enhance image feature by reducing noise devoid of smoothing and losing data by Improved Kaun Filter (IKF).

Segmentation: The whole teeth from an OPG image can be portioned using Improved Artificial Bee Colony (IABC) clustering approach, followed by morphological post-processing performed for classification accuracy.

Feature Extraction: To differentiate teeth and to evaluate ten features that have been eradicated.

Classification: The main aim is to recognize specific person by CA and that is useful for forensic medicine. An effectual procedure specified ENN–DO is used for assessing the age of individual in contrast to Demirjian's age estimation.

1.3.2 Dataset Description

Here, the OPG (i.e. Orthopantomograph) dataset is accumulated from Kovai Scan Centre, Coimbatore. A few OPG teeth images of 4–18-year-old south Indian adolescents who had strong teeth are used for assessing DA and CA.

1.3.3 Pre-Processing Stage

The pre-processing is used to advance an image eminence and to lessen the noise. To eradicate noise and to examine image accuracy, modified Kuan filter has been anticipated in this system, which functions without eliminating edges or image features. Moreover, noise is transformed from multiplicative noise model into an image-based additive noise model. Minimum Mean Square Error (MSE) remains utilized for modeling of original image form. Outcome of pre-processing of gray level value is provided in Equation (1.1):

$$GL_{i,j} = \sum_{s=1}^{m=3,\,n=3} \sum_{t=1}^{m=3,\,n=3} Cp_{s,t} \times W_{s,t} + M_{s,t} \times \left(1 - W_{s,t}\right) \tag{1.1}$$

where (s,t) specifies image pixels and (m,n) specifies rows and columns of pixel data. $Cp_{s,t}$ is depicted as center pixel in filter window; $M_{s,t}$ is specified as mean value of intensity within window; and $W_{s,t}$ is specified as weighting factor. Based on weight, noise reduction has been analyzed. Therefore, to enhance noise reduction and image quality, weight factor is optimized using Random Search Algorithm (RSA). Therefore, function with KF is termed as Modified KF.

In general, RSA is defined using sequence of iteration $\{W_k\}$, $k = 0, 1, \ldots, k$, based on previous pixel points and algorithmic parameter Θ comprises mean vector and covariance matrix. Present iteration W_k specifies collection point. However, iterations are capitalized to specify random variables, showing probabilistic nature of RSA.

Generic search algorithm is based on two significant processes. It generates candidate points and update procedure.

1.3.4   TEETH SEGMENTATION USING IABC

Moreover, IABC clustering is utilized to identify teeth boundaries. This clustering algorithm shows teeth identical pixel regions in one group and normal (value) of all pixels in another group.

In ABC algorithm, D-dimensional is provided as solution space of problem, where D specifies extent of optimization factors. IABC is a choice-based method for candidate neighborhood solutions in Onlooker Bee (OB) stage. The choice-based method is sourced of data shared using Employed Bees (EB). With EBs is for processing standard suitability rate and rate are put away in to memory. In this manner, OB selects the nearest data from memory.

Haphazardly carefully chosen fitness value can be provided in Equation (1.2)

$$avg_m^{popu} = \frac{1}{IJ} \sum_{i=1}^{IJ} suit_i \tag{1.2}$$

where avg_m^{popu} specifies EB's standard suitability rate of population at 'm' iteration and IJ specifies amount of EBs. EB's fitness value are examined with avg_k^{popu} and EBs elucidations, which is superior than avg_k^{popu} is stored to board. Solution duration is evaluated by Equation (1.3)

$$D_i = K.suit_i \tag{1.3}$$

where 'K' specifies optimistic perpetual figure; 'i' specifies suitability rate of ith EBs; D_i specifies waiting time of memory which is proportional to EB's suitability rate. Subsequently, OB's neighbors (x_{kj} in Equation (1.4)) are no longer chosen from memory.

EB's volume and OB are both IJ (swarm of food sources) which is associated to the amount of food sources. For each food source's locality, EB is assigned to it. For EB, whose overall quality is equivalent to food source quality, new sources are acquired based on Equation (1.4):

$$v_{st} = x_{st} + \varphi_{st} \left(x_{st} - x_{kt} \right) \tag{1.4}$$

where $s, k = \{1, 2, ..., IJ\}$, $t = \{1, 2, ..., D\}$; φ specifies randomized actual figure within range of $[-1, 1]$; 'k' specifies index number chosen randomly in Bee colony. Based on production of new solution $v' = \left\{ x_{s1}', x_{s2}', ..., x_{sD}' \right\}$, this solution is in contrast to the original solution $v = \{x_{s1}, x_{s2}, ..., x_{sD}\}$. The acquired solution is superior than the earlier solution; Bee provides new solution or remembers former solution. OB selects food source based on likelihood as in Equation (1.5):

$$P_i = \frac{suit_i}{\sum_{j=1}^{IJ} suit_j} \tag{1.5}$$

where suit$_i\rightarrow$ the fitness of the solution v; IJ specifies location of food sources. Then, OB needs new solution in selected food source mentioned in Equation (1.5), as utilized with EB. With Scout Bee (SB), food source fitness has not been improved for certain trial numbers, which is discarded. This specifies negative feedback in IABC and EB's food source occurs over SB and formulates random search with Equation (1.6):

$$x_{id} = x_d^{min} + r\left(x_d^{max} - x_d^{min}\right)\tag{1.6}$$

where r = a specifies random real number inside range [0, 1]; $x_{min, j}$ and $X_{max, j}$ specify lower and upper border; 'd' specifies problem space of dimension. Fitness value of cluster centers is effectually segmented. IABC clustering pseudo code is explained in the subsequent sections.

1.3.4.1   Fitness Value Approximation

The cluster center of tooth image holds 'G' gray levels [0, ..., G − 1] where random distribution is provided in histogram $h(g)$. So to shorten this fitness value approximation, histogram is normalized with probability distribution function as in Equations (1.7)–(1.9):

$$h(g) = \frac{n_g}{N}, h(g) > 0\tag{1.7}$$

$$N = \sum_{g=0}^{G-1} n_g\tag{1.8}$$

$$\sum_{g=0}^{G-1} h(g) = 1\tag{1.9}$$

where n_g specifies the number of pixels in cluster center with 'g' gray level and 'N' is the entire amount of pixels in the clustered image. Histogram function is the combination of Gaussian probability function as in Equations (1.10) and (1.11):

$$p(x) = \sum_{i=1}^{k} P_i.P_i(x)\tag{1.10}$$

$$p_i(x) = \sum_{i=1}^{k} \frac{P_i}{\sqrt{2\pi\sigma_i}} \exp\left[\frac{-(x-\mu_i)^2}{2\sigma_i^2}\right]\tag{1.11}$$

where P_i is class 'i' probability; $p_i(x)$ is gray level random variable 'x' in class 'i', with μ_i as the mean; σ_i is the standard deviation of ith p.d.f.; and 'K' is the number of classes in image. $\sum_{i=1}^{k} P_i = 1$ is satisfied. Mean square error among histogram function $h(x_i)$ and Gaussian mixture $p(x_i)$ is given in Equation (1.12):

$$J = \frac{1}{n} \sum_{j=1}^{n} \left[p(x_j) - h(x_j) \right]^2 + \omega \left| \left(\sum_{i=1}^{k} P_i \right) - 1 \right| \tag{1.12}$$

In case of IABC optimization, the algorithm lies in stochastic principles. Determination of probability function may lie in point far away and algorithm provides higher ability to locate and carry out global minimum.

ALGORITHM 1:   IABC Clustering-Based Teeth Region Segmentation

Input: image pixels
Output: grouping image pixels
Prepare parameter values. Set 'IJ' threshold value (Population Size), employment bees with %, non-employment bees with 50%, produce Food Number ($IJ/2$) randomly with possible elucidations, maximum amount of repetitions with maxCycle (MCN), stagnation figure is bound (if optimum value does not enhance after repetition, then reorganize possible elucidation)
fitness suit$_i$ of populace is computed in teeth image
for cycle = 1
Repeat
For EB
{
Novel elucidation v_{ij} is generated with Equation (1.4)
}
Probability values P_i are evaluated for solution i using Equation (1.5):
For OB
{
Solution i is chosen based on P_i
Novel solution v_{ij} is generated
Fitness fit$_i$ is evaluated
Greedy selection procedure is used
}
If abounded solution for SB is attained, substitute it with novel solution which is generated randomly by Equation (1.6)
Lower bound $X_{\min, i}$ and upper bound $X_{\max, j}$ shows inside range [0,1]
Finest solution is stored till
For cycle = cycle + 1

Until cycle = MCN

Determine fitness value by approximation

Perform normalization with probability distribution function using Equation (1.8)

Perform histogram function with Gaussian probability function for segmenting clustered dental image

Compute g.p.f. with $p_i(x) = \sum_{i=1}^{k} \frac{P_i}{\sqrt{2\pi\sigma_i}} \exp\left[\frac{-(x-\mu_i)^2}{2\sigma_i^2}\right]$

Mean Square Error attained after segmenting clustered dental image is performed with Equation (1.12)

Stochastic approach to attain global minimum is attained with MSE computation

If g.p.f. lies far away

Then

{

Higher segmentation ability

}

Else

No appropriate segmentation

End if

End

1.3.5 FEATURE EXTRACTION

In detection and classification, features are most important. Features are taken as an input for classification. Then, certain features such as Haralick features, GLCM, Haufsdroff distance, tooth density, distance between crown and root, size. Symmetrical features like concavity, roughness, area, convexity and perimeter has been hauled out.

Gray-Level Co-occurrence Matrix GLCM: It characterizes image texture through pixel pair computation by exact values and enumerated spatial association with an image. Moreover, correlation, energy, homogeneity and contrast are taken out.

Haralick features: These features are utilized to mine essential texture data from co-occurrence matrix. Here, Mean, Contrast, Entropy, Correlation and Sum of squares are hauled out.

Haufsdroff distance: It is used to detect capability among all probable relative position of teeth shape.

Crown length: Distance between Buccal Cement Enamel Junction (CEJ) and buccal cusp tip.

Root length: Distance among root apex and buccal CEJ.

Tooth density: It is evaluated as an average value from the parts such as canines, incisors, molars, premolars of perpetual teeth.

Roughness: It is specified by intensity variance between pixels.
Concavity or Convexity: It is responsive to shape analysis.
Area: Actual amount of pixels in the teeth region.
Perimeter: It measures the distance between teeth borders.

The above-mentioned mined features are utilized for age classification.

1.3.6 DENTAL AGE CLASSIFICATION

DA is categorized with ENN classifier. To enhance the ENN teaching method, DO algorithm has been anticipated. DA-based age classification is in contrast to CA age and shows forecast precision through execution of all classification algorithms.

Features gathered are provided as an ENN classifier input. ENN is a specific feedback neural network type sourced on Back Propagation (BP) neural network's hidden layer that works as delay operator. Therefore, the network system has the ability to adapt time-varying dynamic characteristics and strong global stability. The ENN structure is depicted in Figure 1.3.

There are four layers in the ENN architecture: input layer, hidden layer, output layer and context layer. Context layer is cast off to recognize output of hidden layer, which is considered as step delay operator. With Back Propagation (BP) network, hidden layer output is related to input through storage and delay of context layer. This approach is sensitive to chronological data; the feedback of interior network is able to raise capability to deal with active facts. Internal state has active mapping function and have competency toward adjust time-varying features.

$$I(n) = f\left(w_2 I(n-1) + w_1\left(I(n-1)\right)\right) \qquad (1.13)$$

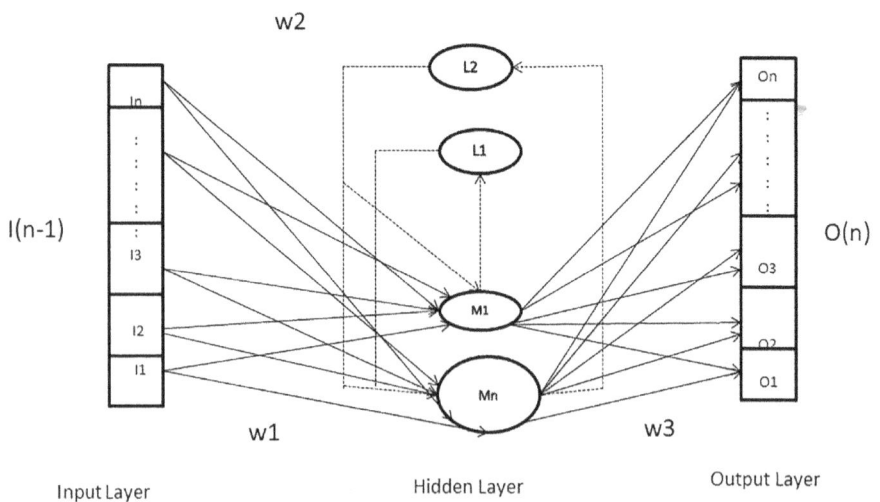

FIGURE 1.3 The ENN architecture.

f is transfer function of hidden layer, which takes sinusoidal function as

$$f(x) = \frac{1}{\left(1 + e^{-x}\right)} \tag{1.14}$$

'p' specifies output layer transfer function, which is a linear function; i.e.,

$$O(n) = g\left(w_3 x(n)\right) \tag{1.15}$$

ENN utilizes BP algorithm to update weights; network error is:

$$E = \sum_{k=1}^{m} \left(a_n - O_n\right)^2 \tag{1.16}$$

where a_n specifies vector object output to enhance ENN prediction accuracy; ENN weight value is optimized with Dragonfly Optimization (DO).

1.3.7 OPTIMAL WEIGHT OF ENN SELECTION USING DO

DO algorithm is cast off to optimize ENN weight and to enhance classification accuracy. Dragonflies possess exclusive swarming features with two functions: migration (dynamic migratory) and hunting (static feeding). An optimization using meta-heuristic has two stages (exploitation and exploration) for dragon flies; migration and hunting are similar to these stages. To show the dragonflies swarming characteristics, three primitive swarming standards are given: alignment (A), separation (S) and cohesion (C). It also considers two ideas: distraction external enemies (E) and attraction in food sources (F) direction to acquire swarm survival rate.

Step and position vectors are measured for dragonfly's location updating the search space and movement's imitation. Dragonfly's location is depicted in step vector, as in Equation (1.17).

$$X_{t+1} = \left(sS_i + aA_i + cC_i + fF_i + eE_i\right) + w\Delta X_t \tag{1.17}$$

where 's' defines weight separation, S_i specifies ith individual separation, 'a' specifies weight alignment, A_i specifies ith individual alignment, c specifies cohesion weight, C_i specifies ith individual cohesion, 'f' specifies food factor (i.e. effectual weight), F_i specifies food source of ith individual, e represents enemy factor, E_i specifies ith individual enemy location, w specifies inertia weight, t specifies iteration counter.

Position vectors are considered as exposed in Equation (1.18):

$$\Delta X_{t+1} = X_t + \Delta X_{t+1} \tag{1.18}$$

where t specifies sum of iteration.

Hence, throughout optimization (i.e. optimal weight selection), with adaptively tuning the swarming parameters, diverse characteristics can be accomplished. As well, neighborhood with definite radius is supposed around every artificial dragonfly. Neighborhood region is amplified in addition to one cluster at ultimate phase of optimization to unite to the global optimum. Based on the above-given procedures, DA is classified in an effectual manner.

1.3.7.1 Threshold Determination

After computation of optimal weight using DO, optimal threshold determination is performed in data classes organized as global optimum. Thus, threshold values are attained through evaluating general likelihood miscalculation of nearby tasks as given in Equation (1.19):

$$E(T_h) = P_{h+1}.E_1(T_h) + P_i.E_2(T_h) \tag{1.19}$$

where $h = 1, 2, ..., k - 1$

$$E_1(T_h) = \int_{-\infty}^{T_h} p_{h+1}(x)dx \tag{1.20}$$

$$E_2(T_h) = \int_{T_h}^{\infty} p_h(x)dx \tag{1.21}$$

Here, $E_1(T_h)$ is the probability of mistakenly classified pixels in $(h + 1)$ class, while $E_2(T_h)$ is the probability of incorrectly classified pixels of hth class. T_h is considered as the threshold value between hth and $h + 1$ class. One of the T_h is selected as $E(T_h)$, which is to be minimized, while differentiating $E(T_h)$ with T_h and equating results to zero. Then, an optimal threshold value to appropriately classify the images with T_h is provided as in Equations (1.22)–(1.25):

$$AT_h^2 + BT_h + C = 0 \tag{1.22}$$

$$A = \sigma_h^2 - \sigma_{h+1}^2 \tag{1.23}$$

$$B = 2.(\mu_h \sigma_{h+1}^2 - \mu_{h+1}\sigma_h^2) \tag{1.24}$$

$$C = (\sigma_h \mu_{h+1})^2 - (\sigma_{h+1}\mu_h)^2 + 2.(\sigma_h\sigma_{h+1})^2 . \ln\left(\frac{\sigma_{h+1}P_h}{\sigma_h P_{h+1}}\right) \tag{1.25}$$

Based on Equations 1.22–1.25, there are two possible solutions, that is, a positive value lies inside the interval. Thereby, it provides a feasible classification outcome.

ALGORITHM 2:   ENN–DO-BASED DA CLASSIFICATION

Inputs: Consider n input, m output;

Number of hidden neurons are r; weight of input layer is w_1; weight of under-take layer is w_2; weight of hidden layer to output layer is w_3; $u(k-1)$ is input of neural network; $x(k)$ is output of hidden layer; $x_c(k)$ is output of undertake layer; and $y(k)$ is output of neural network.

Output: DA classification
Hidden feature parameters $x(k)$ are randomly generated.
Set $x_c(k) = x(k-1)$
Objective function $f(x) = (1 + e^{-x})^{-1}$, where g is transfer function of output layer
Initially, linear function using $y(k) = g(w_3 x(k))$
Calculated weight value of ENN using $E = \sum_{k=1}^{m}(t_k - y_k)^2$ and DO optimization
While (selecting optimal weight)
{ENN phase}
Calculate primitive standards of swarming, separation (S), alignment (A) and cohesion (C)
The best solution of location is attained and computes positive vectors
Teaching factor is calculated
Calculated objective function using separation weight, alignment weight, cohesion weight, food factor, enemy factor, inertia weight
If X_{t+1} is better than X, then $X = X_{t+1}$
End if {ENN phase}

ALGORITHM 3:   DO PHASE WITH THRESHOLD DETERMINATION

Randomly select the position with $\Delta X_{t+1} = X_t + \Delta X_{t+1}$
While $t \rightarrow$ present iteration
End if {DO phase}
Weight optimal results and global optimum is attained by clustering data
Perform threshold computation for classifying two possible solutions
Error probability has to be determined by integrating adjacent function
Probability of mistakenly classified and error classification is attained with Equations 20 and 21
Then compute appropriately classified optimal value using quadratic function with $AT_h^2 + BT_h + C = 0$
Quadratic intervals provide two feasible outcomes
Appropriately classified positive values lie within the range.
Classify dental age based on positive values
Accuracy is calculated.

1.4   OUTCOMES AND DISCLOSURE

Here, the ENN–DO has been utilized for assessing the DA. ENN–DO classifier has been reviewed and compared with the prevailing Demirjian method and also other classification schemes such as FNN-TLBO, MELM-SRC and RBFN. Firstly, OPG teeth images are trained and then tested. The assessment is done by MATLAB. The arranged interaction eventual outcome of proposed ENN–DO is shown in Figure 1.4.

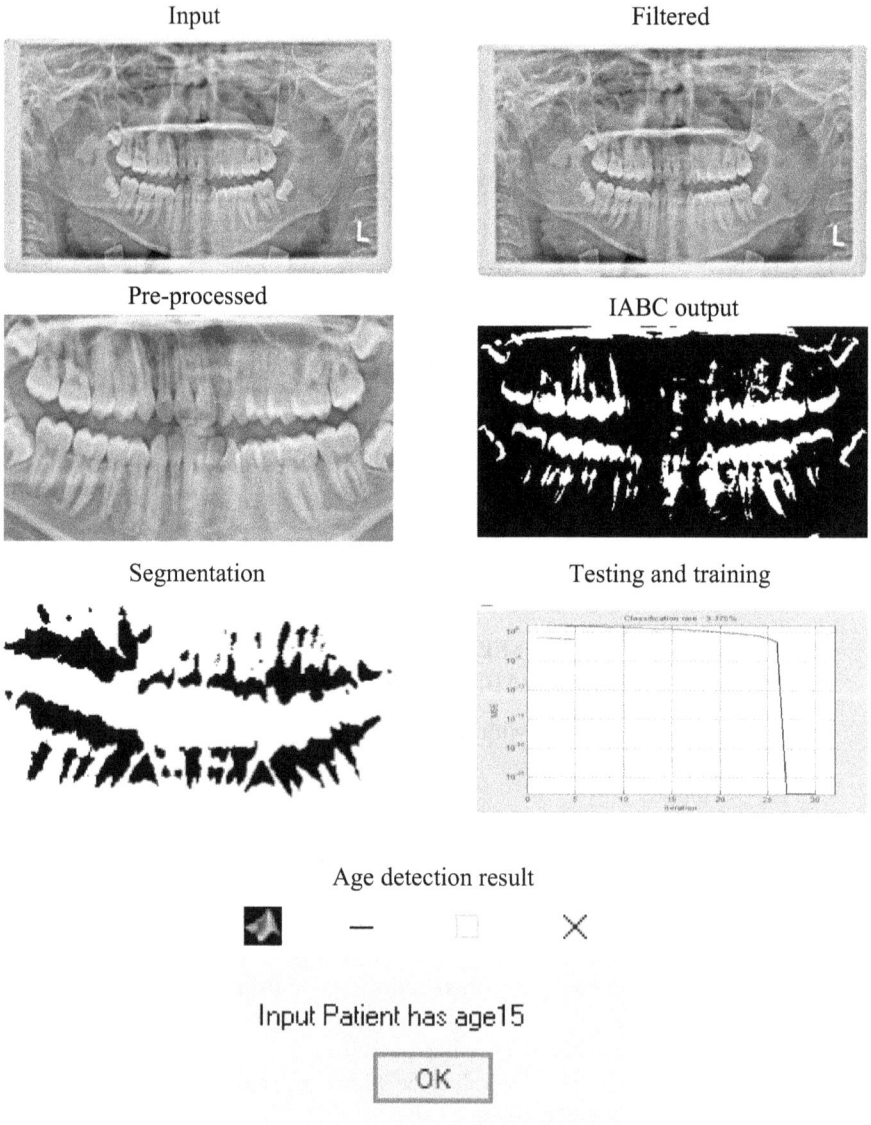

Input

Filtered

Pre-processed

IABC output

Segmentation

Testing and training

Age detection result

Input Patient has age15

OK

FIGURE 1.4   Outcome of input teeth image using anticipated ENN–DO.

The ADF and IKF comparison results are shown in Table 1.1. It illustrates that IKF attained better results compared with ADF in terms of optimization. The performance is measured in terms of PSNR and MSE values. It shows the IKF attained good PSNR compared to ADF as well as less MSE with good optimized result.

The general enactments of entire segmentation schemes are presented in the graph shown in Figure 1.5. In Figure 1.5, the proposed IABC has medium level of attainment show based on exactness rate of 85.5%, specificity rate of 91.05% compared than existing Active Contour Model (ACM) with Analytic Hierarchy Process (AHP) optimization then Active Contour Model (ACM) with Jaya Optimization (JO). Due to the improved process of ABC, the segmentation accuracy has been increased.

The exactness implementation of all DA schemes is represented in the graph shown in Figure 1.6. Inferred from the valuable interest, diminishment and useful teeth region with improved partition then morphological post-processing has been

TABLE 1.1
Performance Measures between ADF and IKF

Performance Measures	ADF	Proposed IKF
MSE	32.1717	31.1939
PSNR	33.2241	35.2272

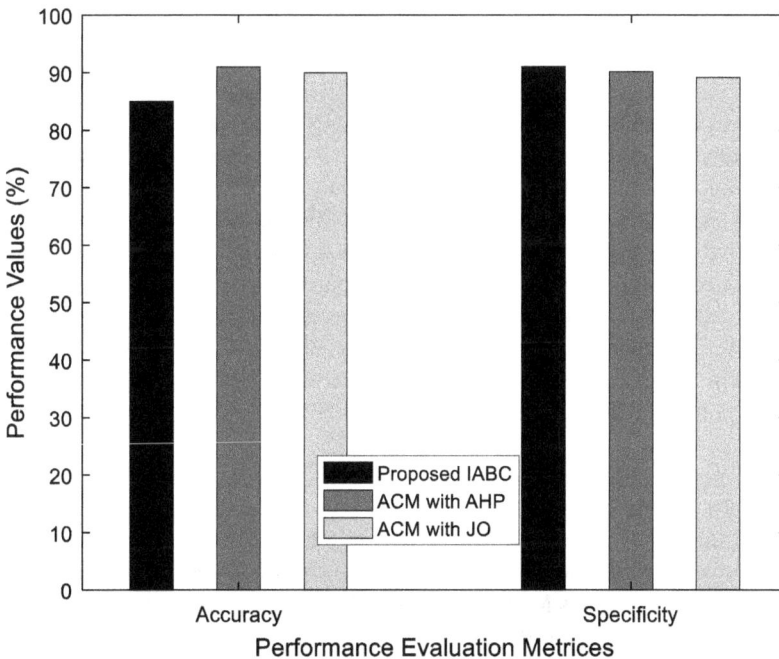

FIGURE 1.5 General enactments for all segmentation schemes.

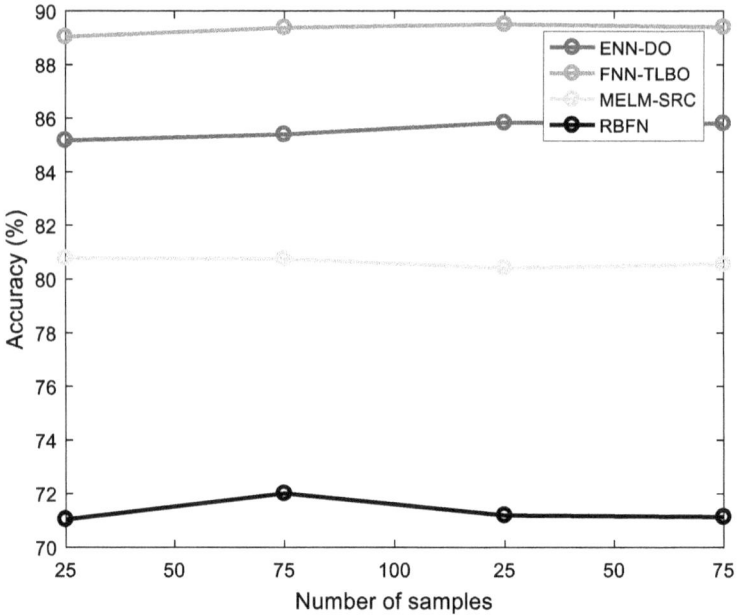

FIGURE 1.6   Contrast of accuracy between diverse DA classification schemes.

upgraded the precision results of the ENN–DO achieve enhanced outcome than MELM-SRC and RBFN.

The specificity performance of all DA schemes is represented in the graph shown in Figure 1.7. It validates the specificity of ENN–DO is −6.87%, 2.11% and 11.26% expanded than FNN-TLBO, MELM-SRC and RBFN. Based on the high positive and negative results, the specificity of anticipated ENN–DO is increased.

The precision execution of all DA schemes is presented in the graph shown in Figure 1.8. It determines the precision of ENN–DO is 0.031%, 19.763% and 25.393% expanded than FNN-TLBO, MELM-SRC and RBFN. Because of high specificity and positive rate, precision of anticipated ENN–DO remains prolonged. ENN consumes reduced computational complexity compared to other classifiers.

The recall implementation of all DA schemes is presented in the graph shown in Figure 1.9. It validates review of ENN–DO is −0.25%, 21.25% and 28.92% contracted than FNN-TLBO and prolonged than MELM-SRC and RBFN. Due to the low error rate and high specificity esteems, ENN–DO is analyzed with different techniques.

The F-measure performance of all DA schemes is presented in the graph shown in Figure 1.10. It proves the review of ENN–DO is −1.69%, 15.77% and 24.96% expanded than FNN-TLBO, MELM-SRC and RBFN. Owing to the high accuracy and investigation rate, the f-measure of anticipated ENN–DO is refined high than existing classifiers such as MELM-SRC and RBFN and low esteem looked at FNN-TLBO. It demonstrates that the proposed ENN–DO outperforms MELM-SRC and RBFN but is lesser than FNN-TLBO (Table 1.2).

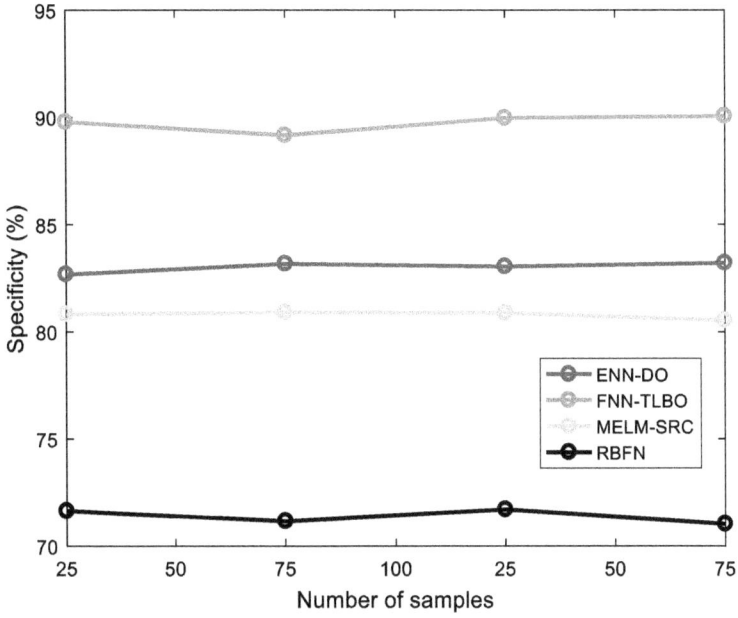

FIGURE 1.7   Contrast of specificity between diverse DA classification schemes.

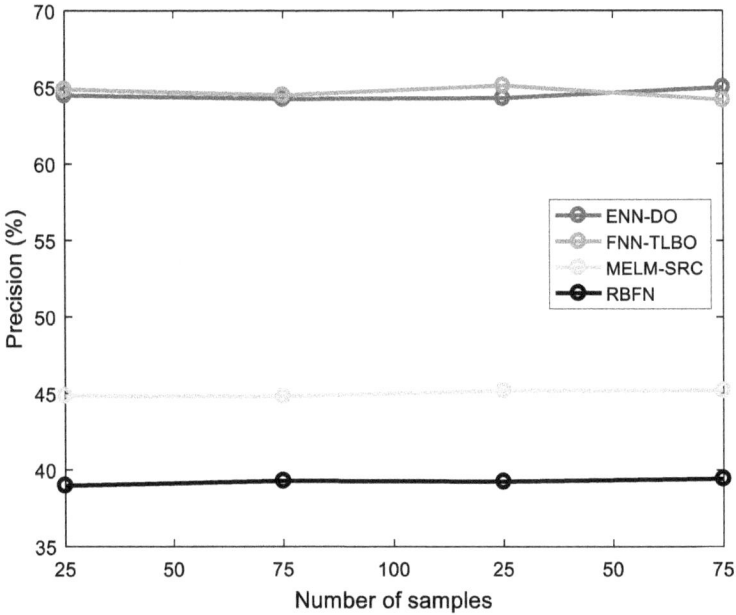

FIGURE 1.8   Contrast of precision between diverse DA classification schemes.

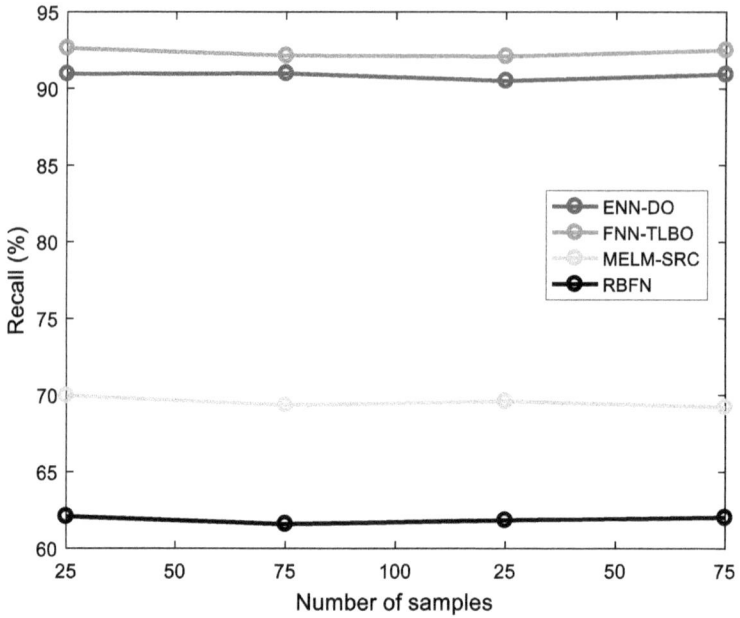

FIGURE 1.9   Contrast of recall between diverse DA classification methods.

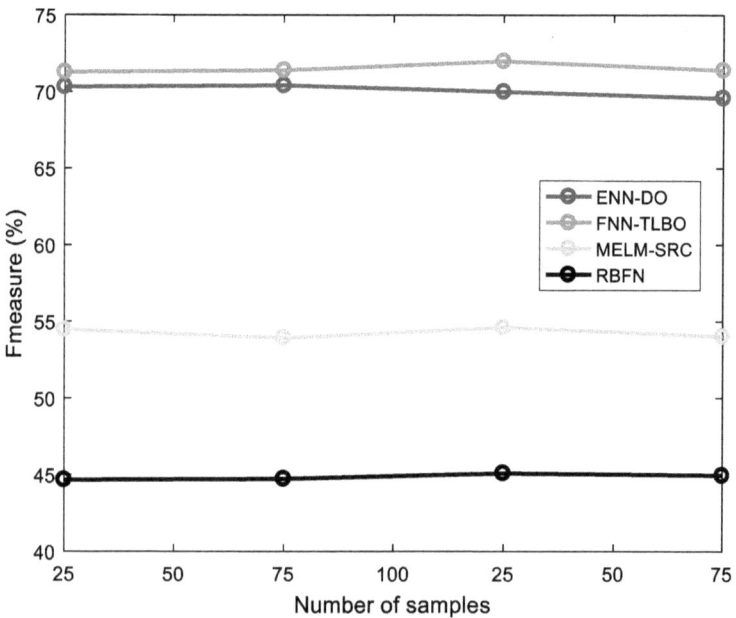

FIGURE 1.10   Contrast of F-measure between diverse DA classification methods.

TABLE 1.2

Total Enactment Mathematical Values for All DA Recognition Schemes

Performance Matrices (%)	Proposed ENN–DO	FNN-TLBO	MELM-SRC	RBFN
Accuracy	85	89	80	72
Specificity	82.25	89.12	80.14	70.99
Precision	64.183	64.152	44.42	38.79
Recall	90.25	92	69	61.33
F-measure	69.43	71.12	53.66	44.47

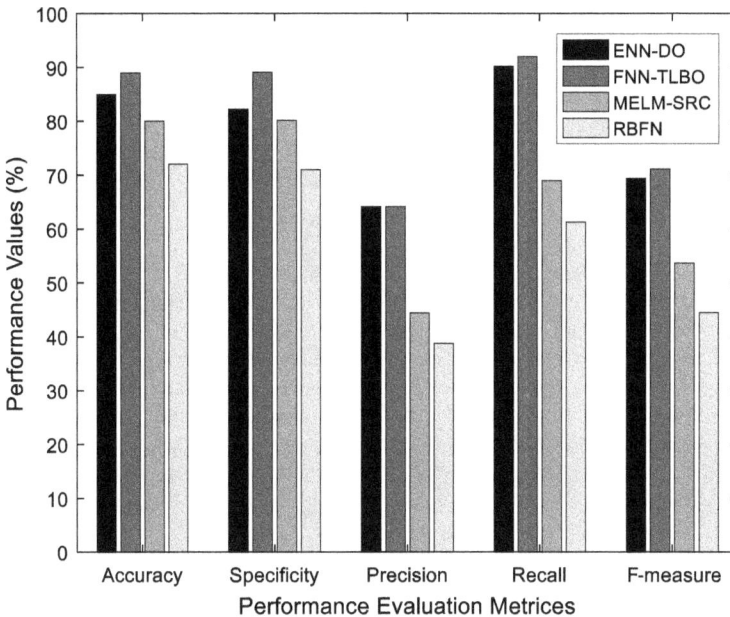

FIGURE 1.11 Expectation of whole enactments for all DA recognition methods.

Overall computation of DA recognition is shown in Figure 1.11. It illustrates that anticipated ENN–DO has acquired superior performance with 85% accuracy, 82.25% of specificity, 64.183% of precision, 90.25% of recall and 69.43% of f-measure in contrast to the prevailing MELM-SRC and RBFN. Due to effectual pre-processing and segmentation, anticipated scheme acquired superior performance than MELM-SRC and RBFN but lower than FNN-TLBO.

1.5 CONCLUSION AND FUTURE WORK

Here, ENN–DO-based DA recognition has been anticipated. The ultimate objective of this recognition is to predict age in an effectual manner in comparison with Demirjian's age estimation for South Indian Populace. Initially, teeth image is pre-processed for lessening noise than improving segmentation. For accuracy enhancement, teeth are portioned for pre-processing with IABC clustering. In this clustering, the teeth region-based pixels are grouped in one and other are grouped as another one. Then, the effectual features are extracted from the teeth region. These pixels are transferred as an input to the ENN classifier. In this ENN, the optimal weight is selected by using the DO scheme. The experimental outcomes demonstrate that anticipated ENN–DO acquired superior accuracy of 85% in contrast to the prevailing algorithms such as RBFN and MELM-SRC. In future, other classification schemes will be evaluated by using large dataset of teeth images and will also be predicted for all ages.

BIBLIOGRAPHY

Arany S, Iino M, Yoshioka N. 2004. Radiographic survey of third molar development in relation to chronological age among Japanese juveniles. *J Forensic Sci*; 49:534–8.

Bunyarit SS, Jayaraman J, Naidu MK, Yuen YR, Danaee M, Nambiar P. 2017. Modified method of dental age estimation of Malay juveniles. *Leg Med (Tokyo)*; 28:45.

Cameriere R, Ferrante L, Cingolani M. 2006. Age estimation in children by measurement of open apices in teeth. *Int J Leg Med*; 120:49–52.

Demirjian A, Goldstein H, Tanner J M. 1973. A new system of dental age assessment. *Hum Biol*; 45:221–7.

Gahleitner A, Nasel C, Schick S, Bernhart T, Mailath G, Dorffner S, et al. 1998. Dental magnetic resonance tomography (dental MRI) as a method for imaging maxillo-mandibular tooth retention structures; *Rofo*; 169:424–8.

Gaudino C, Cosgarea R, Heiland S, Csernus R, Beomonte Zobel B, Pham M, et al. 2011. MR-Imaging of teeth and periodontal apparatus: an experimental study comparing high-resolution MRI with MDCT and CBCT. *Eur Radiol*; 21:2575–83.

Haavikko K. 1970. The formation of alveolar process and clinical eruption of the permanent teeth. An orthopantomographic study. *Proc Fin Dent Soc*; 66:104–70.

Hägg U, Matson L. 1985. Dental maturity as an indicator of chronological age: the accuracy and precision of three methods. *Eur J Orthod*; 7:25–34.

Kapoor AK, Thakur S, Singhal P, Chauhan D, Jayam C. 2017. Compare, evaluate, and estimate chronological age with dental age and skeletal age in 6–14-year-old Himachali children. *Int J Health Allied Sci*; 6(3):143.

Kihara EN, Gichangi P, Liversidge HM, Butt F, Gikenye G 2017. Dental age estimation in a group of Kenyan children using Willems' method: a radiographic study. *Ann Hum Biol*; 44(7):614–21.

Koshy S, Tandon S. 1998. Dental age assessment: the applicability of Demirjian's method in South Indian children. *Forensic Sci Int*; 94:73–85.

Kvaal SI, Kolltveit KM, Thomsen IO, Solheim T. 1995. Age estimation of adults from dental radiographs. *Forensic Sci Int*; 74:175–85.

Liversidge HM, Smith BH, Maber M. 2010. Bias and accuracy of age estimation using developing teeth in 946 children. *Am J Phys Anthropol*; 143:545–54.

Lockemann U, Fuhrmann A, Püschel K, Schmeling A, Geserick G. 2004. Arbeitsgemeinschaft 2004für Forensische Altersdiagnostik der Deutschen Gesellschaft für Rechtsmedizin: Empfehlungen für die Altersdiagnostik bei Jugendlichen und jungen Erwachsenene außerhalb des Strafverfahrens. *Rechtsmedizin*; 14:123–5.

Machado MA, Daruge E, Fernandes MM, Lima IFP, Cericato GO, Franco A, Paranhos LR. 2017. Effectiveness of three age estimation methods based on dental and skeletal development in a sample of young Brazilians. *Arch Oral Biol*; 85:166–71.

Mincer HH, Harris EF, Berryman HE. 1993. The A.B.F.O. study of third molar development and its use as an estimator of chronological age. *J Forensic Sci*; 38:379–90.

Mirjalili S. 2015. Dragonfly algorithm: a new meta-heuristic optimization technique for solving single-objective, discrete, and multi-objective problems. *Neural Comput Applic*; 27(4):1–21.

Nolla CM. 1960. The development of the permanent teeth. *J Dent Child*; 27:254–66.

Nyström M, Haataja J, Kataja M, Evälahati M, Peck L, Kleemola-Kujala E. 1986. Dental maturity in Finnish children, estimated from the development of seven permanent mandibular teeth. *Acta Odontol Scand*; 44:193–8.

Olze A, Bilang D, Schmidt S, Wernecke KD, Geserick G, Schmeling A. 2005. Validation of common classification systems for assessing the mineralization of third molars. *Int J Leg Med*; 119:22–6.

Olze A, Peschke C, Schulz R, Schmeling A. 2008. Studies of the chronological course of wisdom tooth eruption in a German population. *J Forensic Leg Med*; 15:426–9.

Olze A, Reisinger W, Geserick G, Schmeling A. 2006a. Age estimation of unaccompanied minors Part II. Dental aspects. *Forensic Sci Int*; 159:S65–7.

Olze A, Taniguchi M, Schmeling A, Zhu BL, Yamada Y, Maeda H, et al. 2003. Comparative study on the chronology of third molar mineralization in a Japanese and a German population. *Leg Med (Tokyo)*; 5(Suppl 1):S256–60.

Olze A, van Niekerk P, Ishikawa T, Zhu BL, Schulz R, Maeda H, et al. 2007. Comparative study on the effect of ethnicity on wisdom tooth eruption. *Int J Leg Med*; 121:445–8.

Olze A, van Niekerk P, Schmidt S, Wernecke KD, Rosing FW, Geserick G, et al. 2006b. Studies on the progress of third-molar mineralisation in a Black African population. *Homo* 57:209–17.

Orhan K, Ozer L, Orhan AI, Dogan S, Paksoy CS. 2007. Radiographic evaluation of third molar development in relation to chronological age among Turkish children and youth. *Forensic Sci Int*; 165:46–51.

Prieto JL, Barberia E, Ortega R, Magana C. 2005. Evaluation of chronological age based on third molar development in the Spanish population. *Int J Leg Med*; 119:349–54.

Rath H, Rath R, Mahapatra S, Debta T. 2017. Assessment of Demirjian's 8-teeth technique of age estimation and Indian-specific formulas in an East Indian population: A cross-sectional study. *J Forensic Dent Sci*; 9(1):45.

Schmeling A, Geserick G, Kaatsch HJ, Marre B, Reisinger W, Riepert T, et al. 2001. Recommendations for age determinants of living probands in criminal procedures. *Anthropol Anz*; 59:87–91.

Schmeling A, Grundmann C, Fuhrmann A, Kaatsch HJ, Knell B, Ramsthaler F, et al. 2008. Criteria for age estimation in living individuals. *Int J Leg Med*; 122:457–60.

Staaf V, Mörnstad H, Welander U. 1991. Age estimation based on tooth development: a test of reliability and validity. *Scand J Dent Res*; 99:281–6.

Tymofiyeva O, Proff PC, Rottner K, During M, Jakob PM, Richter EJ. 2013. Diagnosis of dental abnormalities in children using 3-dimensional magnetic resonance imaging. *J Oral Maxillofac Surg*; 71(7): 1159–69.

Tymofiyeva O, Rottner K, Jakob PM, Richter EJ, Proff P. 2009. Three-dimensional localization of impacted teeth using magnetic resonance imaging. *Clin Oral Investig*; 14(2): 169–76.

Willems G. 2001. A review of the most commonly used dental age estimation techniques. *J Forensic Odontostomatol*; 19:9–17.

Willems G, Van Olmen A, Spiessens B, Carels C. 2001. Dental age estimation in Belgian children: Demirjian's technique revisited. *J Forensic Sci*; 46(4):893–5.

2 Robotic Assistants from Personalized Care to Disease Diagnosis and Treatment Plans in Healthcare Applications

M. Karthiga, P. Dhivya, and S. S. Nandhini
Bannari Amman Institute of Technology, Erode, India

V. Santhi
PSG College of Technology, Coimbatore, India

CONTENTS

2.1 INTRODUCTION

Artificial intelligence (AI) is a replication of human intelligence, which comprises computer systems or robots, automatically reproducing the reasoning competences that people assistants with dissimilar human minds possess, which include gaining

DOI: 10.1201/9781003181668-2

knowledge and trouble solving. AI is a device that gains deep learning knowledge and axioms that all people look forward to be using these days. Machines acquire knowledge which consists of procedures for different forms of actions, which include regression, clustering, and systems should learn from the data. The extra data given to the algorithm enhance its performance. Deep data acquisition is entirely self-control AI, mainly based on artificial neural networks. Deep understanding of procedures and the need for data is to learn how to complete tasks.

Individual fitness and excellence in life are associated with routine factors, such as workouts, food, sleep, strain, habit, or recreation. Assisted technologies bring interpolations and lifestyle notices all over the day based on an individual's vigorous signs via digital devices. Inside healthcare organizations, AI changes the traditional healthcare systems to operate, optimize, and interact with patients and bring health-care facilities to increase the efficiency of patient care.

Nowadays, AI-based technologies are merged with humans' normal life; the application of AI-based skills will be vital for every single organization. While deep learning has been innovative in difficult problem solving in AI fields for many years, organizations must consider the computational costs involved in training procedures that use large quantities of data, which are combined into our daily lives. Today, many start-ups and intelligence genomics biotechnology corporations in San Francisco are providing healthcare solutions and facilities using skills based on AI. The healthcare industry's most extensively used AI application serves healthcare experts by identifying suitable treatment solutions. It is proposed to merge many schemes with AI competences to improve healthcare applications that might be used to offer medication alerts, patient education materials, and current health metrics. It is pointed out that triggered devices (personal assistant) significantly affect the moni-toring and assistance of patients when medical workers are not available. Smart robots reinforced by AI can also process and increase the work of clinicians with various analysis, treatment methods, reduced costs, and better response time to patient needs [1].

There are countless instances where a personal digital assistant or chatbot could help doctors, nurses, patients, or their families. There are so many possible situations where chatbots can arbitrate and lighten the load of health specialists, who are also used to solve specific problems in the health sector. The personalized chatbot will work to encourage patients by determining test errors and providing data in a respon-sive and efficient manner via email or text. Researchers will also be able to track patients' happiness, cancellations, absences, and successfully finished examinations with the help of application. In some cases, healthcare chatbots can even connect patients with doctors for treatment. These smart procedures for talking or supply messages are used for primary care. If the little medical assistant cannot securely answer the queries raised, it is transferred to a doctor in real life. As the number of health chatbots grows, it is decided to compile a list of the most capable ones to get a sense of the healthcare chatbot industry.

2.2 ROBOTIC ASSISTANCE IN HEALTHCARE

The following robotic assistance is used in healthcare most popularly [2].

2.2.1   ONEREMISSION

For tumor patients and cancer survivors, the application authorizes them by providing a complete list of post cancer diets, movements, and practices, curated by experts in integrative medicine. There is no need to continually trust a doctor. They can research the tumor risks and assistance of a particular food product.

2.2.2   YOUPER

The modern practical examination monitors and improves users' emotional health with quick, modified chats using emotional techniques. To help in extra progress of their expressive health, the app offers modified thoughts and the ability to track attitude and expressive health. As users connect with the chatbots, they will study more about them and refine the knowledge to meet their requirements.

2.2.3   BABYLON HEALTH

In the first case, customers report their disease symptoms to the app, which compares them to a disease database using speech recognition and then recommends a course of action. In the second case, which already exceeds the normal deal of a chatbot, the doctor attends and observes the patient to diagnose the patient and then marks remedies or refers to a professional if necessary.

2.2.4   FLORENCE

The chatbot is essentially a "personal nurse" and works on Facebook Messenger, Skype, or Kik. "She" can repeat patients to take their pills, which could be a useful feature for older patients. Then Florence sends you a chat message every time you need to take the pill. Additionally, Florence can monitor the user's health conditions, such as body weight, mood, or period, helping them achieve their goals. The chatbot also has the ability to discover the adjacent pharmacy or doctor's office in case you need it.

2.2.5   HEALTHILY

This permitted podium proposes actionable health information based on very precise sources and allows the user to make the best choices for their health. It's essentially an AI-based sign checker offered for iOS, Android, Facebook Messenger, Slack, KIK, and Telegram. The medical services provided include pharmacies observation, testing centers booking, doctor's appointment, and rational fitness recommendations.

2.3   USAGE OF AI IN IMPROVING HEALTHCARE SYSTEM

The expansion of robotics and AI continues to provide physicians, helpers, and hospitals with automation and patient care assistance. The robots are used to improve healthcare. Initially there is a plan to turn engineering robots into accuracy machines

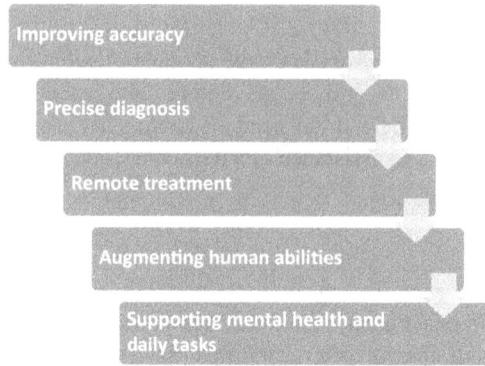

FIGURE 2.1   Effective use of AI.

for operation and beyond. The presence and development of the DaVinci robot and this emblematic video of grape operations show the advancement of technological development. But imposing as it is, robotics are human-controlled systems so far. Nowadays, AI systems surpass the best consultant in merging all the knowledge available in all medical records [3]. It decides that the automated machines substitute qualified physicians and make them more effective in several areas as given in Figure 2.1.

Robotics has no sentiments and will not get tired. If this seems like the perfect doctor of medicine, it's because it's also the brains of numerous robots that have previously been used in the world's greatest hospitals. The human surgeon assumes a subordinate supervisory role as long as the software is appropriately configured for the current surgery. High precision of small microrobots serves and dispenses drugs right away or even performs microsurgery like as blood vessel cleansing.

The true power of AI is in recognizing patterns that can be used to identify a variety of situations. It examines a greater number of cases and hunt for links between hundreds of variables in present checkup research. Mechanical systems have so far been proven to compete with, and even surpass, the greatest doctors in several areas. A Japanese endoscopic technology, for example, detects colon cancer in real time and with high accuracy. However, it pales in comparison to IBM Watson, which has previously achieved a cancer diagnosis accuracy rate of 99%. It came up with the notion of using a robot for remote medical drives in the 1990s, but communication systems at the time couldn't provide the level of care needed to heal military persons in the field. The system stays to support these endeavors, but it appears that mechanical activity actually requires human colleagues for tidiness and different obligations, which indeed makes things more troublesome and not reasonable. The US Department of Defense subsidized examination, which created an autonomous automated consideration framework to treat injured military in out of reach sites. Some clinical robots support patients in adding to clinical staff. For instance, AI can help incapacitated patients walk again and act naturally by making decisions like parental figures. One more use of the innovation is a shrewd prosthesis. It has sensors which are more responsive and exact than the first body parts. By utilizing the capacity to cover and join them to the individual's muscles.

The robots can serve human purposes that serve to the sick or old patients and release them from worries. It makes the patients stay positive, repeat their medications, and realize routine instructions such as temperature, blood pressure, and sugar levels. The individual helpers also have built-in behavior and emotion analysis skills, which are particularly useful for unhappy patients. There is a lot of work in a clinic, and it is not just the doctors who can help. The Moxi robot takes mind of refueling, transporting items, and housework. The great supplementary robot is a UV light sanitizer robot, which enters and leaves a hospital room only when it is free of germs. With the growth of technological advancement, smart healthcare uses a new group of data skills such as the IoT, big data, cloud computing, and AI to alter the old-style system globally to make it more effective, more practical, and more efficient. The idea of smart healthcare is the key skills that provide smart healthcare and then present the current state of system in numerous areas. So, let's look at the problems with system and offer answers [4].

Today's era is that of computerization. With the development of skills and technical theory, old medicine based on biotechnology has started to become digitalized and informed. Smart healthcare that incorporates a new group of information technology has developed. It is a complete and multi-level change, not just a minor technical advancement. This change is personified in the following elements, such as the evolution of the medical model, the evolution of the construction of computerization, changes in organization, and deviations in the concept of avoidance and care. These alterations face the separate desires of people by refining the effectiveness of medical care, which significantly improves the experience of medical and health service employees and signals the direction of the future growth of contemporary medicine. The idea of this system is to identify the key skills and clarify the findings and tests by studying the state of application of these skills among the leading physicians.

The new smart system is made up of several viewers, such as doctors, patients, hospitals, and research institutes. It has multiple directions such as disease detection, decision making, and prevention. From the patient's point of view, he can use portable plans to screen his health at any time, request medical help from virtual assistants, and use remote families to use services remotely from the physician's perspective. A variety of clinical choice care systems are used to aid and recover diagnosis. The impact of AI in healthcare is given in Figure 2.2.

Smart homes also do processes that enhance the breathing skill. The smart families in healthcare fall into home automation and health monitoring. It can deliver humble needs when gathering data and extends help to needy people, reducing dependence on health professionals. It is used to recover their quality of life at home. Patients can know their health disorder on their own through apps and a health data platform. The stress detection and reduction system is the example of portable sensor used in medical field to constantly notice the pressure levels of the human body and reduce the stress. It is also likely to mix health data with portable devices in one system to make a ranked fitness choice care system capable of fully exploiting the data collected for an effective analysis of the disease. It permits to make collaboration among the physicians and patients with other researchers. It permits patients to access the telemedicine guidance, while clinicians can vigorously observe the patient's condition. Physicians are supported by expert peers and researchers. Mobile

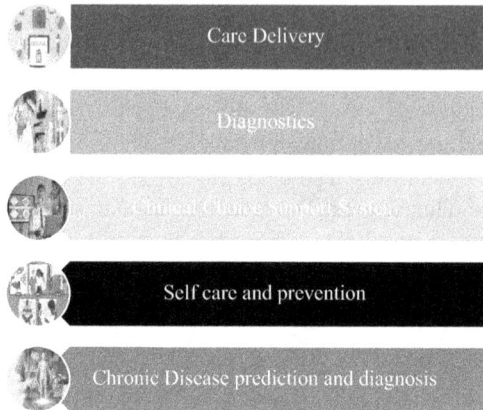

FIGURE 2.2 The impact of AI in healthcare.

technologies such as mHealth decrease errors, decrease the burden of care, recover the correctness of services, and provide a cost-effective platform for services.

Traditional illness risk expectation is based on the creativity of health establishments to collect patient data, evaluate that data with the convincing organization's guidelines, and finally publish the results of the prediction. This method is somewhat delayed in time and does not deliver specific information to others. Predicting risks in smart healthcare is active and modified. It ensures the physicians and patients to participate proactively in monitoring risk with their own surveillance results. The risk forecast model receives data from smart application through cloud network and examines the consequences of large database-based processes to return the expectation to the needy one through messaging system. It helps physicians and patients adapt to the technological environment at all times. It helps in reducing the disease risk.

The replicated subordinate can regularly answer to appropriate data by incorporating patient data, serving medical professionals to attain patients and arrange fitness events. For medical facilities, virtual assistants are used to save manpower and physical resources, and more effectively meet its requirement. It can also be used to establish a dialogue between dissimilar virtual helpers, in particular between overall helpers and extremely qualified supporters. As a result, participants in the medical service's knowledge are greatly improved. Virtual helpers are used to assist in the illness treatment using virtual supporters to expand human psychological health. It can recover the supply of human analysts and provide fitness to more participants in the medical system. Healthcare is made up of three vital mechanisms: local, hospital, and family. The hospital has ICT-based environments particularly those that optimize IoT and automatic procedures to expand care events and present new capabilities. The users must take into account the medical management environment such as the dataset and numerous IoT-based digital schemes attached to digital devices, smart structures, and staffs in hospital management. With the use of advanced technology in scientific research precise practical applications will develop. Automatic transmission of the properties of drugs and targets by AI has dramatically improved the speed of transmission. AI is used to examine and match more cases easily and to monitor for segregation criteria and control the most appropriate target subjects, saving time, and improving the direction of the target population.

A smart system in healthcare can lower the cost of examination, decrease examination time, and overall progress research efficiency. In terms of general decision making, keen healthcare can develop the hospital management system, encourage the path of health enhancement, support the request of expectation strategy, and decrease health costs and social costs. Still, there are a few troubles in the advance system which depends on technical growth and combined hard work of patients, doctors, and healthcare institutions [5].

2.4 NANOROBOTS IN CANCER MEDICATION

Most of the health issues leading to deaths are caused by cancer. Approximately 14.1 million new cases of cancer and 8 million fatalities happened globally in 2012 [6]. By 2030, the prevalence of instances of cancer are expected to rise to 22 million and deaths associated with cancer to 13 million [7]. Due to the remarkable breakthroughs in cancer therapy in recent times, the maturity level of cancer death rate is approximately the same as 50 years ago in the 21st century [8] and the main cause of fatalities associated with cancer has been tumor metastases [9]. Personalized antibody therapy has been a successful and crucial treatment in the last 20 years for the treatment of patients suffering from hematologic cells and structural tumors [10,11]. Globally, more than 25 mono-clonal antibodies have indeed been licensed and new mono-clonal antibodies enter clinical trials at a rate of around 40/year [12]. In medical care, the ubiquitous use of antibodies brings tremendous benefits to healthcare companies. Although antibodies have been unparalleled in treating cancer, many people are still not susceptible or are resistant to antibody-driven therapy [13]. Moreover, the expense of antimicrobial therapy is reasonably high (almost 3,000 dollars for each infusion) and might generate numerous negative effects, occasionally serious ones [14]. Therefore, it is a struggle to recognize patients who can possibly benefit from focused therapy. In order to face this difficulty, the cancer cells of specific patients must be tested and analyzed, which is of significant importance in the new age of customized pinpoint medicine [15].

For the development of human progress, robotics and computational intelligence play an immense role. One of the best examples of this is the automobile sector, where starting from car assembly to delivery, everything is automated in a full and dynamic fashion [16] to improvise production from the beginning of the 1970s [17]. Recently, a lot of advancements in robotic designs have been made in medical sectors in applications like vascular-microtube assembly [18], movement of cell analysis [19], organ-level anatomy [20], and discovery of drugs [21]. Conversely, improvements in robotic design also influence the automation engineering sector as well. Added to that, with the recent requirements for personalized care and medication, personalized robot development has become an opportunistic area in the medical field. In cancer research, drug determination tests (tissue biopsy, tissue culture, removing tumor cells and determining drug therapy, tissue culture, and various biological examinations) are necessary and should be done on all patients individually. The same requires a lot of manpower and several time/days [22]. If these procedures for drug determination tests could be automated, a lot of manpower and time could be consumed, thereby paving the path to personalized treatment procedures for cancer patients. Although this way of introducing automation seems useful, practical implementation of this method is highly challenging.

Substances that are handled is the first problem. Here, monoclonal antibodies stick to the molecular surfaces (like antigen) same as in tumor cells [11], and this can further reduce the tumors through these immune procedures. One such monoclonal antibody is rituximab that was approved in the year 1997 by the United States Federal Drug Administration (USFDA) in the treatment of cancer cells and that could kill the lymphoma cells through three steps [23] such as direct programmed cell death, making cell_lysis, and creating effectors to demolish the tumors. The action of this antibody involves various interactions in the molecular level. For better efficiency of the working of such drugs, determination of target tumor cells which is a crucial activity needs to be addressed. For such automated drug inspection activity, robotic equipments could be involved. The size of this robotic equipment should be very small as it needs to penetrate the body cells. The introduction of nanoautomation [24] techniques facilitates developing of such small nanorobots [24].

The heterogeneous nature of cells is the next issue. In the automobile sector, the automotive robots are designed to perform the same tasks as the parts to be assembled in automobiles are of same shapes and with manual design. Whereas cells are heterogeneous in nature [25] so the automated robots which are designed to penetrate through this cell don't fall under same size category. Besides the dynamic nature of cells, that is, cells undergo lot of changes during their life cycle, the heterogeneous nature of cells is also temporary. This variability in cell nature is also a recent research topic in microbiology [26]. The challenges of the field of life sciences are many. One of these is to understand what components of cellular heterogeneity serve a biological function or contain information that is useful for scientific study. Since cancer cells are constantly changing, micro/nanoautomation technology used in drug susceptibility tests for personalized treatment of cancer patients is difficult to apply. To our knowledge, this issue has not been well addressed from the perspective of automation. While various technologies have emerged over the past periods to increase the automatability of micro- and nanoautomation technologies, additional promises exist for the detection of cancer cells from patients.

2.5 PERSONALIZED TREATMENT PROCEDURES IN CANCER PATIENTS USING NANOAUTOMATION TECHNIQUES

Since humans can't sense, move around, and interact with biological cells at the micro- and nanoscale by themselves, researchers have created robots to extend human capabilities. Since 1959, Nobel Prize-winning physicist Richard P. Feynman has been talking about nanorobots that could be designed and introduced into the human body to perform cellular repairs at the molecular level [27]. As of now, it is hard to create a nanorobot capable of doing a simple medical task by conventional means [28]. However, biological nanomaterials can be used to construct complex "machinery" that can move, sense things, compute information, and make decisions [29]. The findings of Douglas et al. [30] in 2012 showed that nanorobots (35 nm ×

35 nm × 45 nm) can provide drugs directly to target cancer cells, and nanorobots may play a crucial role in other nanotechnology applications. In addition to nanorobots, a big robot can manipulate nanoscale objects, and this type of robot is referred to as a nanomanipulator [31].

Overall, the size of a nanomanipulator is large, though the handling tools, the substances being handled, perception, actuation, and handling exactness are to be in nanolevel [32]. Because of the new conditions (such as fundamental physics, scalability and quantum effects, precision needs) in the nanolevel, nanomanipulator devices are built by employing fresh material, construction, sensing, actuating, controlling, and by connecting new techniques [33]. There are several disadvantages in using SEM (Scanning Electron Microscopy)-based nanomanipulators. First, they require samples to be dried and fixed, so they cannot work with living tissues. Secondly, the ESEM (Environmental Scanning Electron Microscope) handles samples with little moisture, so it does not behave like the conditions inside living cells at normal temperature and pressure. Currently, living cells can't be observed under the ESEM. However, AFM (Atomic Force Microscopy) is able to work in liquids and manipulate single living cells under physiological conditions. The AFM based nanomanipulator can inject drugs [34], cut the cancer cells into smaller pieces [35], indent the surface [36] and visualize the body of a cell [37]. The AFM-based nanomanipulator is predominantly suitable in medical applications due to its precision. Over the past era, noteworthy development has been made in micrometer-sized robots handling of single cells [38]. In 2009, Zhang and others created a swimmable robot equal to a bacterium size. This swimmable robot was developed with a flagellum similar to an artificial bacteria connected to a head made of magnetic material [39]. The movements of these swimmable robots are controlled by electromagnetic devices and so they could be utilized in biological fields. Future challenges include developing the technology for microrobots to steer and move inside the body. A big challenge for researchers lies in identifying these tiny robots after penetrating inside the body and also in knowing how to combine the working of these tiny robots with present diagnostic tools. Along with this, researchers also need to determine a way to control the various armies of these tiny robots inside the body [40]. Micromanipulator is another big robot developed by controlling and manipulating the single cells. These robots are usually part of an array that includes multiple manipulators to control the position and/or orientation of the cell. Micromanipulation has become essential for performing experiments on living systems. Automated robotic systems have been developed for cell capture, immobilization, deposition, patterning, and injection. Additionally, many microrobotic devices have been developed with high-throughput capabilities to study cytoskeleton dynamics [41].

Cancer cells can now be easily detected and various personalized medications can be applied to the cancer patients through micro- or nanorobotic automations. These kinds of advancements are prompted only with the recent achievements in biological applications with these micro- or nanorobotic advancements in the past few decades. The biological samples (such as tissue collection through biopsy, acquiring bone marrow and extracting peripheral samples of blood) are gathered from the cancer patients, and then cancer cells are isolated and given to the pipeline of micro- or nanoautomation technology. These micro- or nanorobots/manipulators undergo sequential cell handling methodologies in the pipeline. In industrial automation, robots in industries are used to carry out a task. In the automotive industry, for example, there is a pipeline that includes different types of industrial robots. The tasks that the industrial robots perform include delivering and injecting drugs onto/into the cell and indenting the cell. Cellular features can be obtained by modeling and analyzing the data. The method starts by recording the quantifiable parameters for each patient's spreadsheet. These parameters are then turned into biological functions, which are used to evaluate drugs and make decisions. After the decision-making process, the doctor uses possible sensitive medicines for clinical treatment.

2.6 ISOLATING CELLS AUTONOMOUSLY IN CANCER MEDICATIONS

To determine the effectiveness of a drug in treating cancer, it is necessary to take samples from patients and test them. The most common way of taking samples is by performing surgery on the patients and extracting tissues. Patients' bone marrow can also be extracted using aspiration, as well as blood. This chapter describes the development of a device that can automatically isolate the cancer cells from strong cells. Such a device would be useful for clinical drug-resistance testing. The first step is to identify physical properties that are different between healthy and cancer cells, such as density or mechanical strength. In the conventional method [42], cells are isolated through fluorescent-activated-cell-sorting (FACS) or magnetic-activated-cell-sorting (MACS). FACS is a process that puts emphasis on fluorescently marked cells using a laser and then applies an electric charge to them based on the dye color. MACS uses magnetic beads to label target cells, and then the beads are passed through a magnetic field that separates non-target cells from the target cells. This method has been used for cell sorting in human stem cell research. Then the target cancer cells are gathered after removing the magnetic effect. FACS and MACS are commonly used in micro/nanoautomation, but they have drawbacks. The main drawback is that the process is time-consuming and dependent on manual labor, which hinders the fidelity of future assays. Microfluidics, a technology that focuses on manipulating liquid at the submillimeter level, has shown promise for detaching cancer cells [43]. Cell sorting methods that use microfluidic devices are faster, less expensive, and more accurate than traditional methods. Microfluidic cell sorting has been used to sort native biological fluids, which eliminates the risk of aerosol-based contamination [44]. The cause of 90% of cancer-related deaths is metastasis, when tumors spread to other parts of the body [45]. Since tumor cells mostly go in blood, the circulating cells of tumor shed into the circulation are an obvious area of interest [46]. In 2010, Circulating Tumor Cells (CTCs) were shown to be effective biomarkers in cancer treatment. In 2016, researchers found evidence that drug sensitivity can also be predicted from the number of CTCs in a patient's blood. Cancer patients have begun collecting their own cells for testing as part of personalized medication [47]. There is very little information about CTCs, and they are usually found in the blood of patients with metastatic cancer. A challenge in isolation is that there are few CTCs in a patient's blood [48]. In 2007, Nagrath et al. [48] created a sole microfluidic platform that could effectively separate CTCs from whole blood samples. The CTC-chip contains a micropost array functionalized with EpCAM antibodies. EpCAM is often overexpressed on carcinomas but is absent from hematologic cells, so anti-EpCAM can provide specificity for capturing CTC from the blood. A study conducted by the Gustave Roussy Cancer Campus observed that the CTC-chip could identify the cancer cells in 99% of patients. The range was 5–1381 cells per milliliter, and approximately 50% of the cells were cancerous. A chip has been developed that captures viable cancer cells (CTCs) directly from whole blood. The chip does not require predilution, pre-labeling, or other processing procedures. By using the chip, CTCs can be isolated quickly and easily. By altering the antibodies attached to the microposts, other rare blood cells can be identified. In 2010, a team of researchers from the same department developed a microfluidics chip for rapid isolation of neutrophils [49]. By using a monoclonal antibody, CD66b-specific mono-clonal antibodies were bound to the external part of microfluidic devices. Neutrophils expressing this CD66b also bound to the external part of the surface. In a recent study, researchers diminished the isolation time and sample volume of blood cells from an hour and milliliters to a few minutes and microliters. This demonstrates the potential part of microfluidics in automatic cell separation. An array of microfluidic silicon nanowire with multiple functionalities nanoparticles is integrated for capturing and detecting CTCs simultaneously and the same was developed by Wany et al. in 2015 [50]. These multiple functional nanoparticles were coated with antibodies like anti-EpCAM for recognizing the cancer cells inside the blood, and these cells after recognition are removed using a magnetic field supplied externally. Further examination or tissue culture is performed on the collected cells from the microfluidic devices.

The procedure for isolating microfluidic cells depends on the exact surface markers of the tumor cells and so the specificity of this procedure is high in [48–50]. But a drawback of this procedure is the difficulty in collecting succeeding cells and minimizing cells' fidelity. Microfluidic procedures without labels for isolating CTCs by using the basic cell properties (like density of the cell, size of the cell, stiffness, and deformability of the cells, compressing nature of the cells, properties of the cell surface, refractive index) have been developed recently by the researchers [51]. Small Reynolds numbers and neglecting inertial effects are in micrometer-sized microfluidic channels. Different size or different shapes or different density cells should be treated in different channels as they respond to inertial effects differently [42]. A novel microfluidic device that is spiral in nature with cross-section that is trapezoidal in nature is developed by Warkiani in 2014 [52] for faster, label-free collection of CTCs from the blood samples. A syringe pump is utilized to pump the collected blood samples through the chip that is spiral in nature. CTCs are gathered nearer to the inner wall because of the inertial effects whereas white blood cells, platelets are collected nearer to the outer wall. The diameters of the CTCs are 3 to 4 times larger than the capillaries of the organs, so they are expected to collect near the inner walls [46]. As CTCs are significantly bigger in size than the normal cells, by utilizing these characteristics of CTCs, in 2014, Huang et al. developed microfluidic devices with varied sizes and with filtration channels to separate this large CTC from the other blood substances [53]. CTCs are present in more than 97% of patients with 19 cells/ millimeter. Whereas specificity of filtration sizes is low because of the overlapping cell sizes of both CTCs and blood cells, and so this is not suitable for all the types of samples [51]. Researchers have explored a label-less isolation of cells methodology using microfluidics by the active application of excitation pressures in recent times. Li et al. designed an acoustic-oriented microfluidic system that can separate CTCs from central blood samples taken from women with breast cancer [54]. Acoustic waves are produced using transducers and the same is mounted on the microfluidic device surface. This acoustic wave could be capable of separating the cells of different properties as each type of cell exhibits different acoustic force with various amplitudes. The major advantage of using acoustic waves is that they are noninvasive and do not affect the function and genetic behavior of the cells. However, this methodology produces less throughput as it involves wavelengths in comparison to channel width in microfluidic device and so the rate at which sorting of different cells are performed is higher [42]. After undergoing various techniques to isolate the cancer cells from the collected samples, the next procedure lies in segregating the cancer cells and providing these cells sequentially to the pipeline of nano- or microautomation techniques for obtaining specific cells. For isolating the cancer cells physically by utilizing a U-shaped barrier, a novel hydrodynamic injection technique was proposed by Di Carlo et al. in 2006 [55]. In order to autonomously distinguish cancer cells for the drug exposure examination, a conveyor-like system is required. Array of different units are accumulated in a chip whereas one chip contains various structures like U-shape formation [55], through-hole [56], microwell [54,57], and micropillar [58]. All these structures are capable of trapping the cancer cells. After pattering the cancer cells, the chip is stacked to a conveyor model wherein a lot of micro- or nanorobots/manipulators are arranged and each of those perform their intended task like identifying the viability of cells, drug injection, displacing the cells, dissecting the cells, and visualizing the various structures of the cells.

For isolating singular cancer cells, a new autonomous conveyor system has to be developed and the same is a hot research study among the scientists nowadays. The next research area is tackling the issues while incorporating the microfluidic devices and the conveyor system, and also the amalgamation of conveyor system and micro- or nanorobots/manipulators. Major drawbacks in microfluidic devices are that they are not reusable as the cells clog and congest together when handled in large numbers. Altogether, microfluidic devices act as a basic building system for autonomous cell gathering and delivery in personalized treatments.

2.7   DATA ANALYSIS

Robotic manipulations facilitate in collecting the features of cancer cells. After collecting the features, the next procedure lies in extracting the biological information from those features. This needs creating a relational model among the features gathered by robots and the natural biological information of the cells. This relational model is injected into a database, where the actual biological information of the cells is automatically mapped to the features. Thus, the prerequisite to be followed here is creating a database for each type of cancer. When the number of patients increases then the size of the database also increases. With the available database, the efficient drugs needed for the cancer patients could be easily recognized. For other medical applications, the samples are required in only small amount to judge the drugs needed for the treatment of the patients. Whereas cancer is highly heterogeneous, so the information gathered from only a small number of samples does not sound good for the treatment of cancer patients. Besides, in the practical treatment of cancer patients, along with the targeted drugs, other chemotherapeutic drugs are also injected into the patients. But the ways to handle the cancer cells with other drugs is still a big research issue in autonomous drug injection examination. Computational intelligence techniques used in various applications are discussed in [59–63].

In the recent era, still researchers are in the process of designing micro- or nanorobots for autonomous biomedical applications. AI techniques are used at the microlevel to build robots to perform applications like micromotors, ultrasonic propulsion system, magnetic propulsion system, and electrophoretic systems. At the nanolevel, biological robots are designed using the molecules. For example, DNAs of the cells are used to design sensors [64] and actin-myosin molecules are designed to act as actuators. The advancement of such micro- or nanorobots design plays a significant role in the improvement of micro- or nanoautomation technique. Another research gap is identified among the label-less physical information of singular cells diagnosed by the nanomanipulators and uncontrolled changes in the functions of cells identified in the clinic. To overcome this issue, the standard procedure of sample preparation in the clinic by the measurement of the robots should be standardized. The threshold values of normal cells and cancer cells should be fixed. Tumor is formed due to the co-evolution of microenvironments related to the cancer cells. Microenvironment acts as an important strategy in determining the cancer cells' behavior. Subsequently, determining the influence of such cancer cells' microenvironments with the effectiveness of clinical drugs is thus another big challenge. Added to it, though the mechanical behavior of the cells for forming the cancer is already evident, the underlying basic mechanisms of the molecules involved in this formation of cancer is still left unknown and the same paves a way to another big research challenge of determining the cellular behavior of cancer growth. For addressing the above-mentioned challenges, the importance of using micro- or nanoautomation techniques in personalized medications is of greater need.

2.8   CONCLUSION

AI and robotic automation have undergone a significant growth in medical industry. Usage of various virtual bots among the patients has now become popular and different specialized hospitals start building their own personalized medications using these computational intelligence techniques. The introduction of micro- or nanorobots for trapping and handling singular cancer cells is a hot research area in biological applications. These techniques are used to identify the personalized drugs essential to treat each and every cancer patient according to their cellular structure. Though the importance of these micro- or nanorobots in cancer treatment is of great significance, developing/designing a complete autonomous robot for singular cancer cell treatment still remains a big challenge. This chapter reveals the progress of some related researches in micro- or nanoautomation growth in this field and also the various technological advancements carried out to improve the personalized medications.

REFERENCES

1. Tian, S., Yang, W., Le Grange, J. M., Wang, P., Huang, W., & Ye, Z. (2019). Smart healthcare: Making medical care more intelligent. *Global Health Journal*, 3(3), 62–65.
2. Jagadeeswari, V., Subramaniyaswamy, V., Logesh, R., & Vijayakumar, V. (2018). A study on medical internet of things and big data in personalized healthcare system. *Health Information Science and Systems*, 6(1), 1–20.
3. Preum, S. M., Munir, S., Ma, M., Yasar, M. S., Stone, D. J., Williams, R., ... Stankovic, J. A. (2021). A review of cognitive assistants for healthcare: Trends, prospects, and future directions. *ACM Computing Surveys (CSUR)*, 53(6), 1–37.
4. Das, S., & Sanyal, M. K. (2020). Application of AI and soft computing in healthcare: A review and speculation. *International Journal of Scientific & Technology Research*, 8(11), 21.
5. Lee, D., & Yoon, S. N. (2021). Application of artificial intelligence-based technologies in the healthcare industry: Opportunities and challenges. *International Journal of Environmental Research and Public Health*, 18(1), 271.
6. Quaquarini, E., D'Ambrosio, D., Gallivanone, F., Hodolic, M., Porta, C., Bernardo, A., & Trifirò, G. (2019, January). Prognostic Value of [18F] Fluorocholine PET Parameters in Metastatic Castrate–Resistant Prostate Cancer Treated with Docetaxel. In *European Congress of Radiology-ECR 2019*, Vienna.
7. Bray, F., Jemal, A., Grey, N., Ferlay, J., & Forman, D. (2012). Global cancer transitions according to the human development index (2008–2030): A population-based study. *The Lancet Oncology*, 13(8), 790–801.
8. Varmus, H. (2006). The new era in cancer research. *Science*, 312 (5777), 1162–1165.
9. Rankin, E. B., & Giaccia, A. J. (2016). Hypoxic control of metastasis. *Science*, 352(6282), 175–180.
10. Scott, A. M., Wolchok, J. D., & Old, L. J. (2012). Antibody therapy of cancer. *Nature Reviews Cancer*, 12(4), 278–287.
11. Weiner, L. M., Murray, J. C., & Shuptrine, C. W. (2012). Antibody-based immunotherapy of cancer. *Cell*, 148(6), 1081–1084.
12. Reichert, J. M. (2010, January). Antibodies to watch in 2010. *Monoclonal Antibodies MAbs*, 2(1), 84–100.
13. Alduaij, W., & Illidge, T. M. (2011). The future of anti-CD20 monoclonal antibodies: Are we making progress?. *Blood, The Journal of the American Society of Hematology*, 117(11), 2993–3001.
14. Maloney, D. G. (2012). Anti-CD20 antibody therapy for B-cell lymphomas. *New England Journal of Medicine*, 366(21), 2008–2016.
15. Collins, F. S., & Varmus, H. (2015). A new initiative on precision medicine. *New England Journal of Medicine*, 372(9), 793–795.
16. Brogårdh, T. (2007). Present and future robot control development—An industrial perspective. *Annual Reviews in Control*, 31(1), 69–79.
17. Hägele, M., Nilsson, K., & Pires, N. J. (2008). *Industrial Robotics*. Springer, Cham.
18. Wang, H., Huang, Q., Shi, Q., Yue, T., Chen, S., Nakajima, M., ... Fukuda, T. (2015). Automated assembly of vascular-like microtube with repetitive single-step contact manipulation. *IEEE Transactions on Biomedical Engineering*, 62(11), 2620–2628.
19. Mach, A. J., Kim, J. H., Arshi, A., Hur, S. C., & Di Carlo, D. (2011). Automated cellular sample preparation using a Centrifuge-on-a-Chip. *Lab on a Chip*, 11(17), 2827–2834.
20. Bhatia, S. N., & Ingber, D. E. (2014). Microfluidic organs-on-chips. *Nature Biotechnology*, 32(8), 760–772.

21. Neužil, P., Giselbrecht, S., Länge, K., Huang, T. J., & Manz, A. (2012). Revisiting lab-on-a-chip technology for drug discovery. *Nature Reviews Drug Discovery*, 11(8), 620–632.
22. Yu, M., Bardia, A., Aceto, N., Bersani, F., Madden, M. W., Donaldson, M. C., … Haber, D. A. (2014). Ex vivo culture of circulating breast tumor cells for individualized testing of drug susceptibility. *Science*, 345(6193), 216–220.
23. Li, M., Liu, L., Xi, N., & Wang, Y. (2016). Applications of atomic force microscopy in exploring drug actions in lymphoma-targeted therapy at the nanoscale. *BioNanoScience*, 6(1), 22–32.
24. Robertson, J. W., Kasianowicz, J. J., & Banerjee, S. (2012). Analytical approaches for studying transporters, channels and porins. *Chemical Reviews*, 112(12), 6227–6249.
25. O'Connor, J. P. (2017, April). Cancer heterogeneity and imaging. *Seminars in Cell & Developmental Biology*, 64, 48–57.
26. Pelkmans, L. (2012). Using cell-to-cell variability—A new era in molecular biology. *Science*, 336(6080), 425–426.
27. Freitas Jr, R. A. (2005). What is nanomedicine? Nanomedicine: Nanotechnology, *Biology and Medicine*, 1(1), 2–9.
28. Nelson, B. J., Kaliakatsos, I. K., & Abbott, J. J. (2010). Microrobots for minimally invasive medicine. *Annual Review of Biomedical Engineering*, 12, 55–85.
29. Lenaghan, S. C., Wang, Y., Xi, N., Fukuda, T., Tarn, T., Hamel, W. R., & Zhang, M. (2013). Grand challenges in bioengineered nanorobotics for cancer therapy. *IEEE Transactions on Biomedical Engineering*, 60(3), 667–673.
30. Douglas, S. M., Bachelet, I., & Church, G. M. (2012). A logic-gated nanorobot for targeted transport of molecular payloads. *Science*, 335(6070), 831–834.
31. Wang, J., & Gao, W. (2012). Nano/microscale motors: Biomedical opportunities and challenges. *ACS Nano*, 6(7), 5745–5751.
32. Sitti, M. (2007). Microscale and nanoscale robotics systems [grand challenges of robotics]. *IEEE Robotics & Automation Magazine*, 14(1), 53–60.
33. Fukuda, T., Arai, F., & Nakajima, M. (2013). *Micro-Nanorobotic Manipulation Systems and Their Applications*. Springer Science & Business Media, Berlin.
34. Guillaume-Gentil, O., Potthoff, E., Ossola, D., Franz, C. M., Zambelli, T., & Vorholt, J. A. (2014). Force-controlled manipulation of single cells: From AFM to FluidFM. *Trends in Biotechnology*, 32(7), 381–388.
35. Yang, R., Song, B., Sun, Z., Lai, K. W. C., Fung, C. K. M., Patterson, K. C., … & Xi, N. (2015). Cellular level robotic surgery: Nanodissection of intermediate filaments in live keratinocytes. *Nanomedicine: Nanotechnology, Biology and Medicine*, 11(1), 137–145.
36. Li, M., Liu, L., Xi, N., Wang, Y., Xiao, X., & Zhang, W. (2014). Nanoscale imaging and mechanical analysis of Fc receptor-mediated macrophage phagocytosis against cancer cells. *Langmuir*, 30(6), 1609–1621.
37. Li, M., Xiao, X., Liu, L., Xi, N., & Wang, Y. (2015). Nanoscale quantifying the effects of targeted drug on chemotherapy in lymphoma treatment using atomic force microscopy. *IEEE Transactions on Biomedical Engineering*, 63(10), 2187–2199.
38. Brand, O., Fedder, G. K., Hierold, C., Korvink, J. G., & Tabata, O. (2015). *Micro-and Nanomanipulation Tools*. John Wiley & Sons.
39. Zhang, L., Abbott, J. J., Dong, L., Kratochvil, B. E., Bell, D., & Nelson, B. J. (2009). Artificial bacterial flagella: Fabrication and magnetic control. *Applied Physics Letters*, 94(6), 064107.
40. Sitti, M. (2009). Voyage of the microrobots. *Nature*, 458(7242), 1121–1122.
41. Lu, Z., Moraes, C., Ye, G., Simmons, C. A., & Sun, Y. (2010). Single cell deposition and patterning with a robotic system. *PLoS One*, 5(10), e13542.

42. Hosic, S., Murthy, S. K., & Koppes, A. N. (2016). Microfluidic sample preparation for single cell analysis. *Analytical Chemistry*, 88(1), 354–380.

43. Sackmann, E. K., Fulton, A. L., & Beebe, D. J. (2014). The present and future role of microfluidics in biomedical research. *Nature*, 507(7491), 181–189.

44. Shields IV, C. W., Reyes, C. D., & López, G. P. (2015). Microfluidic cell sorting: A review of the advances in the separation of cells from debulking to rare cell isolation. *Lab on a Chip*, 15(5), 1230–1249.

45. Spano, D., Heck, C., De Antonellis, P., Christofori, G., & Zollo, M. (2012, June). Molecular networks that regulate cancer metastasis. *Seminars in Cancer Biology, Elsevier*, 22(3), 234–249.

46. Plaks, V., Koopman, C. D., & Werb, Z. (2013). Circulating tumor cells. *Science*, 341(6151), 1186–1188.

47. Danila, D. C., Fleisher, M., & Scher, H. I. (2011). Circulating tumor cells as biomarkers in prostate cancer. *Clinical Cancer Research*, 17(12), 3903–3912.

48. Nagrath, S., Sequist, L. V., Maheswaran, S., Bell, D. W., Irimia, D., Ulkus, L., ... & Toner, M. (2007). Isolation of rare circulating tumour cells in cancer patients by microchip technology. *Nature*, 450(7173), 1235–1239.

49. Kotz, K. T., Xiao, W., Miller-Graziano, C., Qian, W. J., Russom, A., Warner, E. A., ... & Toner, M. (2010). Clinical microfluidics for neutrophil genomics and proteomics. *Nature Medicine*, 16(9), 1042–1047.

50. Wang, C., Ye, M., Cheng, L., Li, R., Zhu, W., Shi, Z., ... & Liu, Z. (2015). Simultaneous isolation and detection of circulating tumor cells with a microfluidic silicon-nanowire-array integrated with magnetic upconversion nanoprobes. *Biomaterials*, 54, 55–62.

51. Chen, Y., Li, P., Huang, P. H., Xie, Y., Mai, J. D., Wang, L., ... & Huang, T. J. (2014). Rare cell isolation and analysis in microfluidics. *Lab on a Chip*, 14(4), 626–645.

52. Warkiani, M. E., Guan, G., Luan, K. B., Lee, W. C., Bhagat, A. A. S., Chaudhuri, P. K., ... & Han, J. (2014). Slanted spiral microfluidics for the ultra-fast, label-free isolation of circulating tumor cells. *Lab on a Chip*, 14(1), 128–137.

53. Huang, T., Jia, C. P., Sun, W. J., Wang, W. T., Zhang, H. L., Cong, H., ... & Zhao, J. L. (2014). Highly sensitive enumeration of circulating tumor cells in lung cancer patients using a size-based filtration microfluidic chip. *Biosensors and Bioelectronics*, 51, 213–218.

54. Wang, Y., Shah, P., Phillips, C., Sims, C. E., & Allbritton, N. L. (2012). Trapping cells on a stretchable microwell array for single-cell analysis. *Analytical and Bioanalytical Chemistry*, 402(3), 1065–1072.

55. Di Carlo, D., Wu, L. Y., & Lee, L. P. (2006). Dynamic single cell culture array. *Lab on a Chip*, 6(11), 1445–1449.

56. Wang, W. H., Liu, X. Y., & Sun, Y. (2008). High-throughput automated injection of individual biological cells. *IEEE Transactions on Automation Science and Engineering*, 6(2), 209–219.

57. Formosa, C., Pillet, F., Schiavone, M., Duval, R. E., Ressier, L., & Dague, E. (2015). Generation of living cell arrays for atomic force microscopy studies. *Nature Protocols*, 10(1), 199–204.

58. Li, M., Liu, L., Xi, N., Wang, Y., Dong, Z., Tabata, O., ... & Zhang, W. (2011). Imaging and measuring the rituximab-induced changes of mechanical properties in B-lymphoma cells using atomic force microscopy. *Biochemical and Biophysical Research Communications, Elsevier*, 404(2), 689–694.

59. Karthiga, M., Sountharrajan, S., Nandhini, S. S., & Kumar, B. S. (2020, May). Machine Learning Based Diagnosis of Alzheimer's Disease. In *International Conference on Image Processing and Capsule Networks* (pp. 607–619). Springer, Cham.

60. Sountharrajan, S., Suganya, E., Karthiga, M., Nandhini, S. S., Vishnupriya, B., & Sathiskumar, B. (2020). On-the-Go Network Establishment of IoT Devices to Meet the Need of Processing Big Data Using Machine Learning Algorithms. In Anandakumar, Haldorai, Arulmurugan Ramu, & Syed, Abdul Rehman Khan (eds) *Business Intelligence for Enterprise Internet of Things* (pp. 151–168). Springer, Cham.
61. Karthiga, M., Sountharrajan, S., Nandhini, S. S., Suganya, E., & Sankarananth, S. (2021, March). A Deep Learning Approach to classify the Honeybee Species and health Identification. In *2021 Seventh International conference on Bio Signals, Images, and Instrumentation (ICBSII)* (pp. 1–7). IEEE.
62. Karthiga, M., Nandhini, S. S., Tharsanee, R. M., Nivaashini, M., & Soundariya, R. S. (2021). Blockchain for Automotive Security and Privacy with Related Use Cases. In Rashmi Agrawal & Neha Gupta (eds) *Transforming Cybersecurity Solutions using Blockchain* (pp. 18–214). Springer, Singapore.
63. Karthiga, M., Sankarananth, S., Sountharrajan, S., Kumar, B. S., & Nandhini, S. S. (2021). Challenges and Opportunities of Big Data Integration in Patient-Centric Healthcare Analytics Using Mobile Networks. In N. Pradeep, Sandeep Kautish, & Sheng-Lung Peng (eds) *Demystifying Big Data, Machine Learning, and Deep Learning for Healthcare Analytics, Elsevier* (pp. 85–108). Academic Press.
64. Hagiya, M., Konagaya, A., Kobayashi, S., Saito, H., & Murata, S. (2014). Molecular robots with sensors and intelligence. *Accounts of Chemical Research* American Chemical Society, 47(6), 1681–1690.

3 A Novel Model for Weather Forecasting Using Deep Learning

S. K. Nivetha, R. C. Suganthe, and C. S. Kanimozhiselvi

Kongu Engineering College, Perundurai, India

N. Senthilkumaran

Vellalar College for Women, Erode, India

Senthil Kumar Muthusamy

University of Technology and Applied Sciences, Sultanate of Oman, Muscat, Oman

S. Ashwini, P. Harinitha, and B. Aishvarya

Kongu Engineering College, Perundurai, India

CONTENTS

3.1 INTRODUCTION

The prediction of atmospheric conditions for a required location and time with the help of science and technology is called weather forecasting. This may have a great impact on environment since it is useful for many farmers and industrial areas.

DOI: 10.1201/9781003181668-3

43

Each year, many parts of the world are destroyed or damaged because of the sudden changes in climatic conditions. To avoid these kinds of damages, the weather is predicted at an earlier stage and precautions are taken to avoid these damages. So, for predicting these, in recent times deep learning methods have been developed at a larger pace.

A. Deep learning

Deep learning techniques are popularly used in various fields like social network filtering, bioinformatics, speech recognition, computer vision, medical image analyzing, audio recognition, board game programs, and material inspection. Compared to human predicted results, deep learning methods produce greater results with more accuracy and performance. Deep learning has several architectures like Recurrent Neural Network (RNN), Convolutional Neural Network (CNN), and Deep Neural Network (DNN). It has now evolved into a major technology that helps in predicting future data. It is used for processing the data and for creating patterns so that the future can be predicted with the past data. It is a subset of artificial intelligence particularly a subset of unsupervised learning and produces results with greater accuracy.

B. Working of deep learning

Deep learning models are built with the inspiration of the human brain. This gives results similar to the conclusion given by a human being after analyzing a given set of data. To obtain this result, a multilayered structure of algorithms called neural networks is used in deep learning. Since large number of inputs may be a time-consuming process when done manually by humans, deep learning models are introduced. This kind of models is used to predict results faster than human beings. The parameters required for processing are to be specified by the programmers themselves. In this model, the input for the next step is taken from the output of the preceding layer. This uses various nonlinear processing units to perform the transformation. Deep learning approaches take a large amount of data and produce results within a short period, whereas humans would take decades to understand and process the data. Many companies have shifted and adapted to these deep learning approaches to save time and to predict the result with greater accuracy. The processing of this is so complicated that it takes decades for humans to understand the process. The datasets that contain data that are neither classified nor labeled are called unsupervised data. The training of machines using these data is called unsupervised learning (UL). The main advantage of the deep learning approach is that it can be applied to UL. This is a major advantage because labeled data are lesser than unlabeled data. Hence deep learning techniques are used in various areas for predicting the future data with more accuracy. Figure 3.1 explains the general process of deep learning.

The task of deep learning is UL. This algorithm uses untagged data to learn patterns. In supervised learning the data are tagged by programmers, whereas UL uses

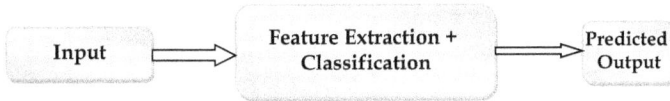

FIGURE 3.1   Deep learning approach.

untagged data. Self-supervision is a process of training UL models with the help of supervised learning methods. These are trained with multiple inputs given in the dataset. The accuracy of the result increases as the number of inputs used for training is increased.

3.2   LITERATURE REVIEW

In the existing works, many statistical modeling techniques including machine learning have been reported to solve the weather forecasting problem. Salman et al. (2018) proposed a Long Short-Term Memory (LSTM) model in which the LSTM memory block is added with intermediate variables signal and pressure. Kim et al. (2017) proposed convolutional LSTM (ConvLSTM) for the prediction of rainfall using weather radar data. LSTM cell with convolution operation is known as ConvLSTM. It is a variant of LSTM.

Troncoso et al. (2015) proposed different types of regression trees (RT) for wind speed prediction. In this work, short-term wind speed is predicted by using eight kinds of novel regression tree structures and very short-term wind speed prediction in real time is done successfully. Whenever the measuring tower collects the new wind speed data, the regression trees allow the retraining of the algorithm because regression trees have a very small computation time.

Sharaff and Roy (2018) made a comparative analysis between nonlinear method and regression methods. To predict the temperature, they analyzed the difference between the performance of linear and nonlinear models. As a result, BPN is suggested for temperature prediction with better results.

Yonekura et al. (2018), proposed a very short-term weather forecast method. They proposed deep learning architectures for rainfall prediction. They considered point prediction model and tenser prediction model. They concluded that DNNs had the highest accuracy for rain prediction.

Roesch and Günther (2019) considered the metrological attributes such as temperature, pressure, and wind speed. They proposed a method that uses recurrent CNN. The complete outputs of training data and testing data are made available by the user.

Broni-Bedaiko et al. (2019) proposed a method that uses LSTM model to forecast ENSO phenomenon. LSTM neural network model has great potential to forecast weather with more data samples.

Xingjian et al. (2015) proposed Convolutional LSTM Network approach for precipitation forecasting. The authors predicted the future rainfall intensity and that is for a local region but for a short period of time. They proposed a new extension called convolutional LSTM and formulated a precipitation prediction so as to address spatiotemporal sequence forecasting problem. In the same year, Salman et al. (2015) compared the prediction performance of RNN, Conditional Restricted Boltzmann Machine (CRBM), and Convolutional Network (CN) models. Frobenius norm is used in each model to evaluate the forecasting accuracy.

Singh et al. (2019) proposed Random forest-based classification for weather prediction. The data are collected using sensors like DHT11 and BMP180 for every 1 hour. The result is predicted and is represented as a confusion matrix.

Ehsan et al. (2019) proposed two error models for wind speed prediction. The two models are Bayesian Additive Regression Trees (BART) and the Quantile Regression Forests (QRF). The result of comparison is based on the point estimates and prediction intervals. They presented that error modeling based on QRF is giving better results than BART in terms of point estimate and predicted results.

Seven machine learning algorithms like Genetic Programming, Support Vector Regression, Radial Basis Neural Networks, M5 Rules, M5 Model trees, k-Nearest neighbors, and Markov chain were used by Cramer et al. (2017) for rainfall prediction. The detection of predictive accuracy and correlations between different climates is also done in this work. The accumulated rainfall is predicted in this model. Two main things discussed in this work are, testing the accumulation of rainfall using several machine learning techniques and testing whether varying climates affects the combination of the data accumulation and algorithms.

Basha et al. (2020) proposed methods based on multilayer perceptron and autoencoders for the rainfall prediction. An autoencoder performs the feature extraction, so it is applicable in time series forecasting. The authors presented a detailed study on several methodologies that are applied for forecasting applications. Hewage et al. (2020) proposed a weather forecasting model using Temporal Convolutional Neural (TCN) Network and LSTM. They concluded that these models are performing better in weather prediction than machine learning models. The authors extended their work (Hewage et al. 2021) by accessing the TCN and LSTM models with two different regressions and achieved accurate predictions up to 12 hours. Karevan and Suykens (2020) proposed transductive LSTM to predict weather condition, and it uses weighted quadratic cost function.

Rodrigues et al. (2018) proposed a novel method based on DNNs to achieve a high-resolution representation from low-resolution prediction mainly considering weather forecasting as a case study. The authors considered supervised learning approach in order to do automatic labeling of data. Both linear regression and NN architecture are used for weather forecasting.

Booz et al. (2019) have done a study using weather data to train the deep learning model. In this study, the authors concluded that there is relationship between the data volume and accuracy obtained. Also, there is a relationship between the data recency and accuracy obtained. Zhou et al. (2019) proposed a CNN model was then built to classify thunderstorm, heavy rain and hail. The performance comparison between CNN model and other traditional methods are also mentioned in this work. The authors concluded that deep learning algorithms are better in providing higher classification accuracy on heavy rain and hail than the other classical machine learning algorithms.

3.3 PROPOSED SYSTEM

The LSTM algorithm is a special kind of RNN, which is commonly used for time series data. To learn long-term dependencies was the main application of LSTM.

The proposed model is based on LSTM networks for which the general model is mentioned in Figure 3.2 and it uses temporal weather data and produces weather predictions. Figure 3.2 represents that, when inputs are given, it processes the data and produces certain output which is stored in the memory cell and is again given as input to the next level for further processing until the desired output is obtained.

This is an unsupervised model and is focused on providing better results than the previously proposed model. In this model we have used many parameters for detecting the atmospheric conditions like temperature, humidity, pressure, wind speed, rainfall, and precipitation. These parameters were chosen based on their importance in increasing the accuracy of the prediction.

3.3.1 LONG SHORT-TERM MEMORY

LSTM is a special kind of RNN. The input is given with the value which is derived as output of the previous step in RNN. RNN provides more accurate predictions from the recent information but words stored in long-term memory cannot be predicted by RNN. This problem of long-term dependencies is tackled by LSTM. If the length of the gap increases, then the performance of RNN decreases. LSTM is used mainly to retain the information for a long period of time. LSTM algorithm is used for processing, predicting, and classifying time series data.

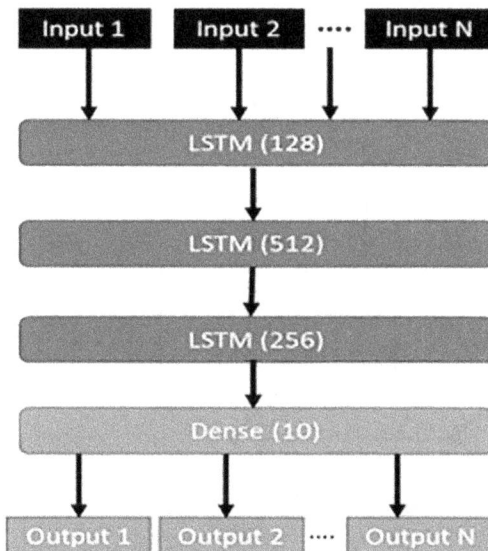

FIGURE 3.2 General LSTM model.

One of the hardest problems to be solved in the data science industry is sequence prediction problems. Sequence prediction problems include predicting sales, understanding movie plots, finding patterns in stock markets data, language translations, recognizing your way of speech, and predicting your next word on your iPhone's keyboard. LSTM provides an effective solution to all the above-mentioned sequence prediction problems.

In RNN there is no information like it's 'important' and 'not so important' because if we want to add a piece of new information, then the existing information is transformed entirely.

This is done by applying a function. It is standard feed-forward neural networks. But, in LSTM the information is not transformed completely because it selectively remembers or forgets things. The LSTM adds information or modifies by additions and multiplications. The information here flows through cell states. So, the information is not modified entirely.

The information in cell state has three different dependencies as shown in Figure 3.3.

1. The previous cell state. After the previous time step, the information present in the memory is stored in this state.
2. The previous hidden state. This state is like the output of the previous cell.
3. The input at the current time step. Here, the new information is provided at that moment.

LSTM processes entire sequences of data and not only the single data points. It has feedback connections. The unit of LSTM is composed of a cell which is one of the units of LSTM that remembers the values over arbitrary time intervals, an input gate,

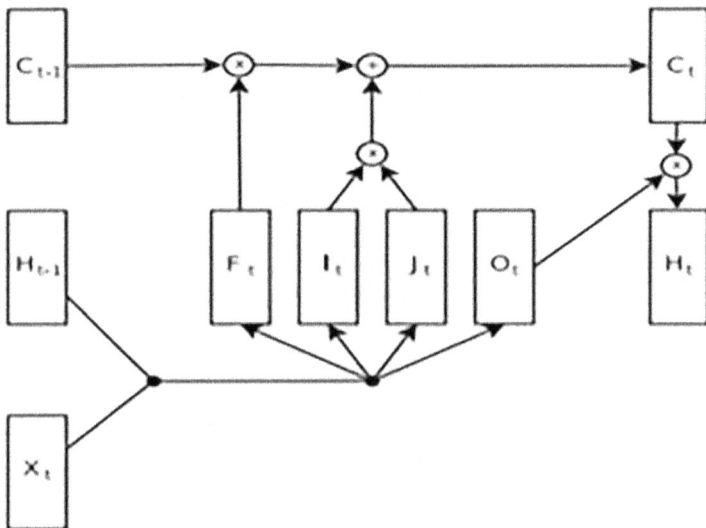

FIGURE 3.3   LSTM memory cell.

an output gate, and a forget gate. The three gates arrange the flow of information into and out of the cell.

LSTM networks are significant for processing, classifying, and making predictions based on time series data. They were developed mainly to solve the problems of exploding and vanishing the gradient that can arise while training with traditional RNNs.

3.3.2 ARCHITECTURE OF LSTM

In Figure 3.4, the LSTM architecture is presented and in that the LSTM unit is composed of a cell, i.e., the memory part of the LSTM and the three 'regulators' called gates. The gates manipulate the memory block which is responsible for 16 remembering things. The flow of information inside the LSTM is achieved by the input, output, and forget gate. The cell state and the hidden state are the states which are transferred to the next cell.

The variables are as follows:

- X_t – input vector to the LSTM unit.
- F_t – forget gate's activation vector.
- I_t – input or the update gate's activation vector.
- O_t – output gate's activation vector.
- H_t – hidden state vector or the output vector of the LSTM unit
- C_t – cell state vector.

Some LSTM units do not have these input, output, and forget gates. For example, gated recurrent units (GRUs) do not have an output gate.

Cell: It is used to keep track of the dependencies between the elements in the input sequence.

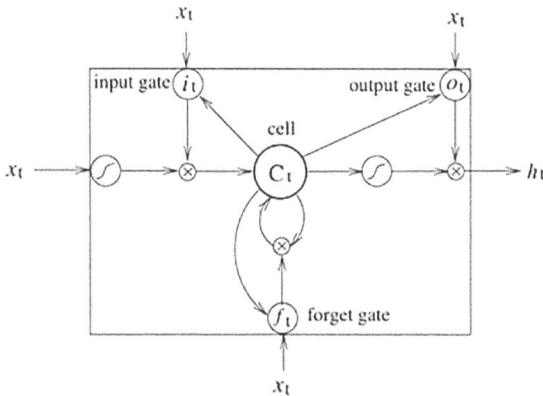

FIGURE 3.4   Architecture of LSTM.

Forget gate: Forget gate is used to remove information from the cell state. Through the multiplication of filters, the information that is no longer important is removed. This gate is used to enhance the LSTM network performance.

Input gate: It is used to add information to the cell state. After completion of some steps, we must ensure that only the important and not redundant information is added to the cell state.

Output gate: The information that runs along the cell state may or may not fit for being output at a certain time. The output gate displays the useful information from the cell state as the output.

The filter is being applied to the cell state vector after it is built on the hidden state values and the input.

3.3.3 Additional Hyper-Parameters

The dropout layer should accompany every LSTM layer. The dropout layer helps to prevent overfitting. This is done during training by ignoring the randomly selected neurons;and by preventing overfitting, the sensitivity to the specific weights of individual neurons is reduced. In order to retain model accuracy and also to prevent overfitting, the dropout value is considered as 20% for each iteration/epoch (Roesch & Günther 2019).

1. Bidirectional: The information derived from the earlier sequences can be used for the latter sequence and vice versa.
2. LSTM (x3) layers: The 30 units here indicate the 30-day windowed historical datasets. The return sequence parameter specifies that the output of each unit is used as the input of the next layer.
3. Dropout ($p = 0.2$): It is used to avoid overfitting of data.
4. Dense layer: Afully connected layer with fiveprediction neurons, and each neuron represents a day into the future

3.4   RESULTS AND DISCUSSION

3.4.1 Overview of Dataset

The dataset used in this work is collected from Kaggle. The dataset has attributes like date, humidity, temperature, pressure, wind speed, rainfall, and precipitation. Various features of the climate are predicted for the next 5 days based on the previous input data.

Dataset 1:
The data belongs to the country Turkey. The dataset comprises 31,200 records. In Table 3.1 the description about dataset used is presented which consists of the number of training and testing samples that are used.

TABLE 3.1
Dataset-1 Description

Dataset	No. of Training Samples	No. of Testing Samples	Total No. of Samples
Weather dataset (Turkey)	28,080	3,120	31,200

TABLE 3.2
Dataset-2 Description

Dataset	No. of Training Samples	No. of Testing Samples	Total No. of Samples
Weather dataset (New Delhi)	54,000	6,000	60,000

Dataset 2:
The data belongs to the state Delhi. The dataset comprises 60,000 records. In Table 3.2 the description about dataset used is presented which consists of the number of training and testing samples that are used.

3.4.2 RESULTS AND DISCUSSION

Models that have been proposed earlier in the literature for weather prediction predict either one two parameters only, for example, predicting only rainfall that too for minimum time limit like up to 24 hrs (i.e. one day forecasting only) or predicting only temperature or both. In our work, we considered four parameters like wind speed, temperature, pressure, and humidity. We also predicted the weather parameters for the next consecutive 5 days. We obtained this result by applying LSTM algorithm. We have used Python-based modules to implement this system. Thus, the Figures 3.5–3.8 show the results of various parameters like wind speed, temperature, pressure, and humidity, respectively, for dataset 1.

The predicted values are the values obtained by applying the LSTM algorithm and the real values are the values that are present in the database. The values achieved as results as per the Figures 3.5–3.8 are also mentioned in Tables 3.3–3.6, respectively. Weather is a natural process, so accurate prediction is critical. Here, we have obtained this result in 50, 150, and 250 epochs, respectively. So, there is a mismatch of values for certain days.

Based on the values given in Table 3.3, it is evident that only for Day 4 it is closely predicting the wind speed at 250 epochs. This may be because of the drastic climatic difference during that period.

If we further increase the number of epochs, we may achieve better accuracy than this but the response time greatly increases. This may be considered as a future work to improve the accuracy with reduced response time.

50 Epoch

```
Predicted wind_speed [[1.839235  1.8582299 1.8656487 1.912636  1.9492133]]
Real wind_speed [3.4 0.5 1.  1.5 0.5]
```

150 Epoch

```
Predicted wind_speed [[1.5409276 1.4639056 1.4322126 1.466282  1.5258117]]
Real wind_speed [3.4 0.5 1.  1.5 0.5]
```

250 Epoch

```
Predicted wind_speed [[1.9046172 1.7857068 1.7086136 1.6366227 1.5860906]]
Real wind_speed [3.4 0.5 1.  1.5 0.5]
```

FIGURE 3.5   Predicted vs. real wind speed (mph).

50 Epoch

```
Predicted temperature [[9.271861 8.952031 8.714903 8.587166 8.581047]]
Real temperature [9.45 9.43 8.75 8.51 8.6 ]
```

150 Epoch

```
Predicted temperature [[9.051559 8.592226 8.180081 7.890688 7.743372]]
Real temperature [9.45 9.43 8.75 8.51 8.6 ]
```

250 Epoch

```
Predicted temperature [[9.366301 9.148042 8.909732 8.670079 8.4915  ]]
Real temperature [9.45 9.43 8.75 8.51 8.6 ]
```

FIGURE 3.6   Predicted vs. real temperature (°C).

Based on the values given in Table 3.4, it is evident that for all the 5 days the predicted temperature is almost accurate at 250 epochs.

Based on the values given in Table 3.5, it is evident that for all the 5 days the pressure is predicted with a closer value.

Based on the values given in Table 3.6, it is evident that for Day 3 it is predicting the humidity much closer than the other days at 250 epochs and for Day2 and Day4 it is predicting a closer value at 50 epochs itself.

Similarly Tables 3.7–3.10 shows the results of various parameters like wind speed, temperature, pressure, and humidity, respectively, for dataset 2.

50 Epoch

```
Predicted pressure [[1006.2678  1006.27026 1006.2803  1006.2743  1006.27747]]
Real pressure [1011. 1011. 1011. 1010. 1010.]
```

150 Epoch

```
Predicted pressure [[1006.1214  1006.11865 1006.11676 1006.11584 1006.1146 ]]
Real pressure [1011. 1011. 1011. 1010. 1010.]
```

250 Epoch

```
Predicted pressure [[1006.45935 1006.4562  1006.45386 1006.4524  1006.4508 ]]
Real pressure [1011. 1011. 1011. 1010. 1010.]
```

FIGURE 3.7   Predicted vs. real pressure (Pa).

50 Epoch

```
Predicted humidity [[81.93046  81.46386  81.203865 80.25712  79.50435 ]]
Real humidity [87. 81. 87. 81. 76.]
```

150 Epoch

```
Predicted humidity [[81.05283 81.62248 82.03764 82.40245 82.13303]]
Real humidity [87. 81. 87. 81. 76.]
```

250 Epoch

```
Predicted humidity [[84.70704  85.082886 85.4966   84.13511  83.04794 ]]
Real humidity [87. 81. 87. 81. 76.]
```

FIGURE 3.8   Predicted vs. real humidity (g/kg).

TABLE 3.3
The Predicted and Real Wind Speed for the Next Consecutive Five Days

	Predicted Wind Speed (mph)			Real Wind Speed (mph)
	50 Epoch	**150 Epoch**	**250 Epoch**	
Day 1	1.839235	1.5409276	1.9046172	3.4
Day 2	1.8582299	1.4639056	1.7857068	0.5
Day 3	1.8656487	1.4322126	1.7086136	1
Day 4	1.912636	1.466282	1.6366227	1.5
Day 5	1.9492133	1.525117	1.5860906	0.5

TABLE 3.4

The Predicted and Real Temperature for the Next Consecutive Five Days

	Predicted Temperature (in Celcius)			Real Temperature
	50 Epoch	150 Epoch	250 Epoch	(in Celcius)
Day 1	9.271861	9.051559	9.366301	9.45
Day 2	8.952031	8.592226	9.148042	9.43
Day 3	8.714903	8.180081	8.909732	8.75
Day 4	8.587166	7.890688	8.670079	8.51
Day 5	8.581047	7.743372	8.4915	8.6

TABLE 3.5

The Predicted and Real Pressure for the Next Consecutive Five Days

	Predicted Pressure (mb)			
	50 Epoch	150 Epoch	250 Epoch	Real Pressure (mb)
Day 1	1006.2678	1006.1214	1006.45935	1011
Day 2	1006.27026	1006.11865	1006.4562	1011
Day 3	1006.2803	1006.11676	1006.45386	1011
Day 4	1006.2743	1006.11584	1006.4524	1010
Day 5	1006.27747	1006.1146	1006.4508	1010

TABLE 3.6

The Predicted and Real Humidity for the Next Consecutive Five Days

	Predicted Humidity (gkg^{-1})			Real Humidity
	50 Epoch	150 Epoch	250 Epoch	(gkg^{-1})
Day 1	81.93046	81.05283	84.70704	87
Day 2	81.46386	81.62248	85.082886	81
Day 3	81.203865	82.03764	85.4966	87
Day 4	80.25712	82.40245	84.13511	81
Day 5	79.50435	82.13303	83.04794	76

TABLE 3.7

Average Humidity Values – Predicted vs. Real Humidity

Number of Epochs	Humidity Values
50	Predicted humidity [[41. 37. 34. 43. 55.]]
	Real humidity [47. 36. 38. 49. 61.]
100	Predicted humidity [43. 37. 33. 45. 58.]
	Real humidity [47. 36. 38. 49. 61.]
250	Predicted humidity [47. 37. 36. 48. 59.]
	Real humidity [47. 36. 38. 49. 61.]

TABLE 3.8
Average Pressure Values – Predicted vs. Real

Number of Epochs	Pressure Values
50	Predicted pressure [[1007. 1006. 1006. 1006. 1007.]]
	Real pressure [1010. 1007. 1005. 1006. 1007.]
100	Predicted pressure [[1008. 1006. 1006. 1007. 1007.]]
	Real pressure [1010. 1007. 1005. 1006. 1008.]
250	Predicted pressure [[1009. 1007. 1006. 1006. 1007.]]
	Real pressure [1010. 1007. 1005. 1006. 1008.]

TABLE 3.9
Average Temperature Values – Predicted vs. Real

Number of Epochs	Temperature Values
50	Predicted temperature [[28. 31. 33. 32. 30.]]
	Real temperature [30. 34. 34. 30. 27.]
100	Predicted temperature [[28. 32. 34. 31. 30.]]
	Real temperature [30. 34. 34. 30. 27.]
250	Predicted temperature [[29. 33. 34. 31. 29.]]
	Real temperature [30. 34. 34. 30. 27.]

TABLE 3.10
Average Wind Speed Values – Predicted vs. Real

Number of Epochs	Wind Speed Values
50	Predicted wind speed [[28. 16. 12. 12. 12.]]
	Real wind speed [24.1 14.8 11.1 13. 11.1]
100	Predicted wind speed [[23. 16. 11. 12. 12.]]
	Real wind speed [24.1 14.8 11.1 13. 11.1]
250	Predicted wind speed [[23. 15. 11. 12. 11.]]
	Real wind speed [24.1 14.8 11.1 13. 11.1]

3.5 CONCLUSION AND FUTURE WORK

The prediction of weather has been done using the LSTM model. This is done using various parameters present in the two different datasets which belong to the country Turkey and the city New Delhi in India. The predicted results give the humidity, pressure, wind speed, and temperature data for the next five consecutive days. It helps to detect the weather for upcoming days and so it helps the people who depend on weather conditions for their daily activities. Depending on the deviations in the climatic conditions, the people may change the activity according to it.

In future, we will try to predict weather for Tamil Nadu and develop a web application based on this ideology to help the user to predict the weather. Also, some additional features like precipitation and solar radiation will be used to predict the weather to still increase the accuracy of the predicted values. Thus, our research work would be very useful to those whose work depends on the atmospheric weather and will also help in giving prior information to the seashore people about vast disasters so that they could be moved to some safer place.

This will be particularly valuable for all farmers and related enterprises as it will assist them inplanning their business early. Also, this will be helpful to the people residing near the seashore to be priorly informed about the changes in atmospheric pressure and to help them shift to some safer place.

REFERENCES

Basha, C. Z., Bhavana, N., Bhavya, P., & Sowmya, V. (2020, July). Rainfall prediction using machine learning & deep learning techniques. In *2020 International Conference on Electronics and Sustainable Communication Systems (ICESC)* (pp. 92–97). IEEE.

Booz, J., Yu, W., Xu, G., Griffith, D., & Golmie, N. (2019, February). A deep learning-based weather forecast system for data volume and recency analysis. In *2019 International Conference on Computing, Networking and Communications (ICNC)* (pp. 697–701). IEEE.

Broni-Bedaiko, C., Katsriku, F. A., Unemi, T., Atsumi, M., Abdulai, J. D., Shinomiya, N., & Owusu, E. (2019). El Niño-Southern Oscillation forecasting using complex networks analysis of LSTM neural networks. *Artificial Life and Robotics*, 24(4), 445–451.

Cramer, S., Kampouridis, M., Freitas, A. A., & Alexandridis, A. K. (2017). An extensive evaluation of seven machine learning methods for rainfall prediction in weather derivatives. *Expert Systems with Applications*, 85, 169–181.

Ehsan, B. M. A., Begum, F., Ilham, S. J., & Khan, R. S. (2019). Advanced wind speed prediction using convective weather variables through machine learning application. *Applied Computing and Geosciences*, 1, 100002.

Hewage, P., Behera, A., Trovati, M., Pereira, E., Ghahremani, M., Palmieri, F., & Liu, Y. (2020). Temporal convolutional neural (TCN) network for an effective weather forecasting using time-series data from the local weather station. *Soft Computing*, 24(21), 16453–16482.

Hewage, P., Trovati, M., Pereira, E., & Behera, A. (2021). Deep learning-based effective fine-grained weather forecasting model. *Pattern Analysis and Applications*, 24(1), 343–366.

Karevan, Z., & Suykens, J. A. (2020). Transductive LSTM for time-series prediction: An application to weather forecasting. *Neural Networks*, 125, 1–9.

Kim, S., Hong, S., Joh, M., & Song, S. K. (2017). Deeprain: ConvLSTM network for precipitation prediction using multichannel radar data. arXiv preprint arXiv:1711.02316.

Rodrigues, E. R., Oliveira, I., Cunha, R., & Netto, M. (2018, October). DeepDownscale: A deep learning strategy for high-resolution weather forecast. In *2018 IEEE 14th International Conference on e-Science (e-Science)* (pp. 415–422). IEEE.

Roesch, I., & Günther, T. (2019, February). Visualization of neural network predictions for weather forecasting. *Computer graphics forum*, 38(1), 209–220.

Salman, A. G., Kanigoro, B., & Heryadi, Y. (2015, October). Weather forecasting using deep learning techniques. In *2015 International Conference on Advanced Computer Science and Information Systems (ICACSIS)* (pp. 281–285). IEEE.

Salman, A. G., Heryadi, Y., Abdurahman, E., & Suparta, W. (2018). Single layer & multi-layer long short-term memory (LSTM) model with intermediate variables for weather forecasting. *Procedia Computer Science*, *135*, 89–98.

Sharaff, A., & Roy, S. R. (2018, May). Comparative analysis of temperature prediction using regression methods and back propagation neural network. In *2018 2nd International Conference on Trends in Electronics and Informatics (ICOEI)* (pp. 739–742). IEEE.

Singh, N., Chaturvedi, S., & Akhter, S. (2019, March). Weather forecasting using machine learning algorithm. In *2019 International Conference on Signal Processing and Communication (ICSC)* (pp. 171–174). IEEE.

Troncoso, A., Salcedo-Sanz, S., Casanova-Mateo, C., Riquelme, J. C., & Prieto, L. (2015). Local models-based regression trees for very short-term wind speed prediction. *Renewable Energy*, *81*, 589–598.

Xingjian, S. H. I., Chen, Z., Wang, H., Yeung, D. Y., Wong, W. K., & Woo, W. C. (2015). Convolutional LSTM network: A machine learning approach for precipitation nowcasting. In C. Cortes, N. Lawrence, D. Lee, M. Sugiyama, R. Garnett (eds.), *Advances in neural information processing systems* (pp. 802–810). Curran Associates, Inc., Red Hook, NY. ISBN: 9781510825024

Yonekura, K., Hattori, H., & Suzuki, T. (2018, December). Short-term local weather forecast using dense weather station by deep neural network. In *2018 IEEE International Conference on Big Data (Big Data)* (pp. 1683–1690). IEEE.

Zhou, K., Zheng, Y., Li, B., Dong, W., & Zhang, X. (2019). Forecasting different types of convective weather: A deep learning approach. *Journal of Meteorological Research*, *33*(5), 797–809.

4 Application of Artificial Intelligence Algorithms for Robot Development

R. M. Tharsanee, R. S. Soundariya,
A. Saran Kumar, and V. Praveen
Bannari Amman Institute of Technology, Erode, India

CONTENTS

DOI: 10.1201/9781003181668-4

4.1 ARTIFICIAL INTELLIGENCE

Artificial Intelligence (AI) has become a necessity for every part of human life in today's world. AI was first propounded by John McCarthy in the year 1956. AI is found to be one of the most important technologies in the field of computer science. AI is a vast discipline which comprises various disciplines like psychology, biology, Machine Learning and human intelligence. AI is a branch of computer which involves the creation of intelligent machines by understanding the behavior of humans. Therefore, AI creates a machine capable of doing human works. Figure 4.1 shows the various fields that are integrated with AI technology (see Figure 4.1).

The history of AI traces back to 1950 when Alan Turing put forth the question 'Will machines think?' In the year 1950, a test named the 'Turing test' was done to test the ability of machines to think intelligently as humans or think differently [1]. The test requires some parameters to be set in the areas of reasoning, automation, semantics and knowledge representation. As the test became successful, thereafter the term AI was introduced by John McCarthy in the year 1956. In 1970s, symbolic AI was the first evolved type of AI which uses human knowledge representation to build an AI system. After that, an agent was discovered in the 1980s which is an

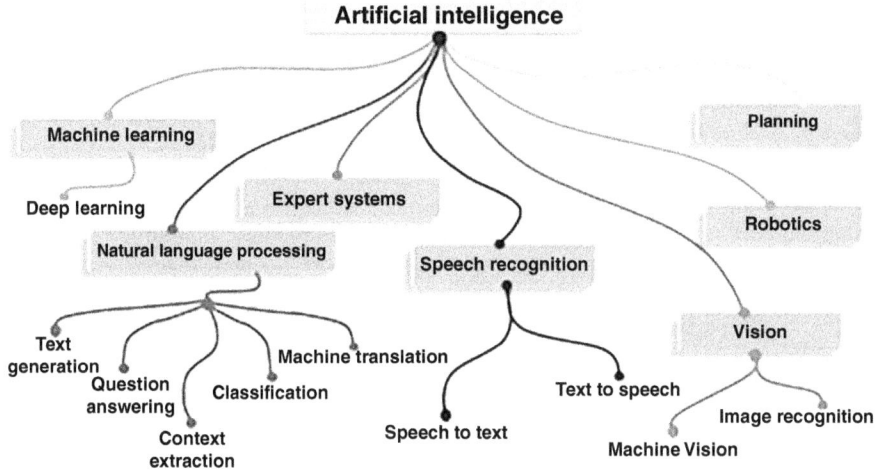

FIGURE 4.1   Relation of AI with other fields.

intelligent system which analyzes its environment and behaves correctly and intelligently. The main functions that can be performed by AI systems include planning, organizing, decision making and task execution. Many scientists predicted that the customer request can be processed by machines without human intervention by using AI techniques. AI is developing every day. The applications of AI are increasing rapidly in a variety of fields like medicine, Robotics, automation and so on [2].

In general, AI can be divided into three categories namely: Narrow AI, general AI and Artificial superintelligence. At present, only Narrow AI has been reached. Due to the evolving nature of Machine Learning and intelligent algorithms, the other two types of AI are under development. Narrow AI is based on two theories in general: The first theory is fearfulness of an empyrean future where intelligent machines will lead the world and make the human beings a slave or distinct. The other theory is an optimistic theory where humans and intelligent systems will work together in order to increase the human life experience.

AI technology/systems are divided into three types based on their abilities to perform the tasks of human:

- Narrow AI: perform narrow list of activities/capabilities of human
- General AI: perform the activities in accordance with the human
- Artificial superintelligence: perform activities that cannot be done by humans.

i.   Narrow AI
     Narrow AI is also called weak AI. It is the only type of AI we have recognized so far. This weak AI can perform only one task that is designed to be and is mostly goal oriented. It is called a weak AI because it performs under specific conditions and restrictions. This AI cannot perform the tasks as that of humans rather it only stimulates the activities of

humans. The Narrow AI system is based on the use of Natural Language Processing (NLP) in order to perform the activities. Narrow AI is coded to work with human beings by understanding the text and speech in natural language.

Narrow AI can be of reactive type or limited memory type. Reactive Narrow AI has no memory/storage capabilities, and it works in a way to emulate the activities of humans without any prior experience to solve a problem. Limited memory AI can have data storage capabilities which can be used to store historical data in order to make decisions and perform tasks.

The applications of Narrow AI include Siri assistant in iPhone, Alexa voice assistant in Amazon, facial recognition system, autonomous cars and robots, spam filters and marketing recommendations [3].

ii. General AI

General AI is also called strong AI. It is a type of system which performs the activities as that of humans. It has the ability to learn and apply suitable methodologies to solve a specific problem. These AI systems have the ability to think, learn new strategies and solve the problem in an intelligent manner. These AI systems use the mind AI theory framework in order to manage the problem situation. This framework has the ability to find the needs, analyze the environment and make a suitable decision to solve the problem. This theory is not a simulation or replication of humans, rather than training the machines to make decisions as humans. The main aim of strong AI is to create machines with human intelligence or to make a machine that is different from the human mind. Due to the rapidly changing capability of the human mind, it is very difficult to create strong AI systems.

One of the world's fastest supercomputers named as Fujitsu-built K is an application of strong AI which would take 40 seconds to simulate a single second of human neural activity. Currently, there is not even a single exact application of strong AI systems due to the difficulties involved in the recognition of human activity [4].

iii. Artificial Superintelligence (ASI)

ASI is a speculative AI which does not imitate humans rather than analyze the problem and the situation and take corrective actions accordingly. ASI has the ability greater than that of human beings in solving the given problems. ASI will create robots that can perform actions more than that of humans and will make humans slave to AI systems. ASI systems will do the activity that can be performed by humans namely art, music, sports, emotional relationship, math's calculation and problem solving. ASI has large memory capacity which can store and process information and respond to stimuli in order to solve the problem. As ASI systems will be capable of outperforming the human being, it is dangerous to make such systems as it would make the human community to become slaves to machines [5].

4.1.1   Use Cases of Artificial Intelligence

4.1.1.1   Infectious Disease Diagnosis Using Machine Learning

Covid-19 crisis affected humans a lot in the past one and half years and many of them were not aware of Covid-19 destruction. The medical experts and government are in need of knowing the future pandemics in advance to take remedial action. AI's and other technologies are widely preferred to predict or detect the future pandemics across the globe [6].

4.1.1.2   Surgical Robots with Intelligence

In the 21st century, the population and the demand for doctors are increasing gradually. With this situation, the robot surgeons are a must, and it can help the surgeon to achieve the new level of precision.

4.1.1.3   Healthcare Applications

The healthcare applications are available in the app store or play store. With these kinds of applications, the patients' health conditions are monitored on a daily basis powered by voice assistance technology. AI guides the patients to take necessary actions to heal whenever they are in need of assistance.

4.1.1.4   Fraud Detection

Banking and Financial sectors are highly affected with fraud transactions since humans aren't able to keep tracking the millions of transactions. To resolve this, the Machine Learning algorithms are effectively used to identify the fraud transactions by checking the thousands of data points at the same time.

4.1.1.5   Investment Prediction

One of the valuable predictive tasks is to identify the better investment plan since Machine Learning helps to predict the advanced market growth, and based on the insights generated, the investment plan can be made by the managers. The goal is to identify the advanced market plan compared to the traditional investment structure.

4.1.2   AI Applications in Manufacturing Sectors

Most of the companies are encouraged to use the concept of AI in manufacturing the products and in supply chain management areas. It is observed or proven that the AI has produced good results in terms of performance tests, packing the finished products and also the handling of materials. Deep Learning algorithms are effectively used in manufacturing of cars, and this shift makes the robots decide which parts have to be picked and how to pick the parts.

4.1.3   AI Applications in Quality Control and Automotive Insurance

The quality control section in automobile industries mainly focuses on checking the painted car bodies. This is a sensitive one and humans can make mistakes in identifying that. The crack detection is also an important one under the quality control process. Machine Learning acts as a substitute for humans to predict all these types of problems. The insurance companies can avail the help of AI applications to claim the insurance for accidents and other unavoidable circumstances. AI applications are more connected with the car, passenger and driver to avoid any cyber threats.

4.1.3.1 Personalized Learning and Smart Learning

The traditional method focuses on providing the same kind of knowledge to all the learners. AI-based learning can identify the knowledge gap of the learners and enhance the learners' level by suggesting the learning recommendations for each and every learner. Apart from the learning recommendation system, AI delivers the digitized content. The benefit of using AI is that visualization and simulation contents and much more learning material types like audios, charts, e-books, videos and so on can be created.

4.1.3.2 Reducing Admin/Staff Work and Increased Accessibility

Some of the tasks can be repeatedly done by the administrative staff on a daily basis. The repeated tasks are grading, correcting the assignments, conducting the examinations or assessments that consume more time. AI can take up all this work and reduce the burden of teachers. With this, teachers can focus on the student's performance and also the need of the student to learn the new skills. The learning boundaries are restricted in the physical classroom learning.

4.1.3.3 Product Pricing and Advertising

Product price fixing is one of the tedious tasks in the retail industry. The market price of all the products has to be known before going to fix the price on it. AI can be used to identify the place where the business advertisement should be. It will improve the businesses by not wasting the money for ineffective marketing.

4.1.3.4 Surveillance and Logistics

The US Pentagon has a proposal named project Maven which incorporates AI algorithms alongside the computer vision to cross check the recorded footage from the unaware serial vehicles.

From the logistics perspective, the autonomous information systems with real-time data are extracted by the Air Force. All the sensors are embedded with the engines and other related systems. The data is fed into the predictive algorithms in order to identify the kind of technicians required to check the aircraft.

4.1.3.5 Client Sentiment Analysis

To enhance the company performance, customer sentiment analysis is a crucial one for processing the information. The company can identify the positive and negative impacts about the likeliness of the product from the customer perspective. At present there are lots of digitized tools that can identify the customer's changed behavior, and these data are collectively used by the AI to predict the customer expectation and also it would be a great one to improve the company's performance.

4.1.3.6   Search Optimization

There is a lot of information available online, but the relevant data has to be recommended first based on the customer search. With the help of AI, the recommendations and the search results are quick and more accurate. For example, instead of typing the movie name, anyone can upload the movie image and easily identify the movie information from the web.

4.2   ROBOTICS

Robotics involves the combination of various fields of engineering and technology in order to create a successful robot or a robotic system. Robotics integrates the various fields of computer science like Machine Learning, AI, embedded systems and computer vision. These systems combine several algorithms at various levels in order to create successful robots. During the 1960s industrial revolution drove the need for the introduction of industrial robots in order to replace human workers. Thereafter, various types of robots have emerged in diverse fields to serve the requirements in the world.

Robots can be defined as an independent device which is capable of performing various actions and can act independently or can be autonomously programmed to perform various activities in order to serve the current industrial needs or human needs. Robots are also called agents. An agent is capable of performing actions by analyzing the environment and using some action rules to solve the problem.

Basically there are two design issues involved in the design of robots or an agent, namely action and perception. The robotic system structure may take various forms of design architectures like agent architecture, subsumption architecture and cognitive architecture. The perception of Robotics involves the common areas of AI and robots. The design of robotic systems is perceived by visual perception systems. In order to perform a specific task, a robotic system must be equipped with various ranges of analyzing capabilities in order to classify and solve the problems.

The main components of a robot include arm or actuator, effector, sensor, controller and locomotive device. A robot arm also called a robot actuator is a series of joints which resembles the human arms. This arm is used to make interaction with the outside effectors. This can also be used to perform activities like painting, welding and serving. The effector is attached to the other end of the actuator which can perform the tasks as that of human arm and fingers. The locomotive device is actually a motor which can be used to provide power for the working of the robots. There are three types of power that can be provided by locomotor, namely electric, hydraulic (liquid power) and pneumatic (air/gas power). The controller is used to perform and monitor the tasks of the robot. The function of the controller is similar to that of a human brain. It also controls the robot at various levels. Sensors are used to sense the environmental conditions in order to make decisions and perform the task correctly. Figure 4.2 shows the general robot structure (see Figure 4.2).

FIGURE 4.2   General robot.

4.2.1   TYPES OF ROBOT

Robots can be of any size, shape and perform different tasks based on their requirements. Each and every robot has its own characteristics. Some of the different types of robots with their characteristics are listed in Table 4.1.

TABLE 4.1
Types of Robot

Characteristics/ Robot Type	Humanoid Robots	Autonomous Robots	Industrial Robots	Swarm Robotics	Cobots
Mode of operation	Human-like activity	Simple task	Complex activity	Mission critical activity	Combination of simple and complex tasks
Control	Autonomous	Self-control and autonomous	Non autonomous	Streamline and autonomous	Autonomous
Adaptability	High	High	Low	High	High
Domain	Unknown	Known/ Unknown	Unknown	Known	Known
Example	Boston Dynamics' Atlas	Hospitality bots	Spider robots	Micro-drones	Sawyer cobot arm

4.2.2 ROBOTIC APPLICATIONS

Since the application of robotic technology is applied in every field and in everyday life of human being, some of those application areas are listed as follows:

i. Healthcare: Healthcare robots are very helpful for continuous monitoring of patients and also help medical practitioners to give the best treatment for the patients in case of a medical emergency. These robots can also be used for drug supply to the patients in case of non-availability of doctors.

ii. Surgery: Robots can be used to perform surgery for patients in case of critical circumstances in which humans cannot perform the surgery. These robots can be programmed to perform a range of services to patients [7].

iii. Education: In the field of education, robots can be utilized to teach students in a lively manner. These robots can also be used to monitor the activity of students and help educators to give good quality of services to the students [8].

iv. Agriculture: In the agriculture sector, robots can be used to monitor the plants thereby helping the farmers to water the crops at an appropriate time. It can also be used to identify the various insects and pests which may cause damage to the plants, thereby reducing the yield of crops. Also these agri-robots can be utilized to apply fertilizers for the crops at various points of time [9].

v. Home automation: Robots can be used to perform various activities at home like cooking, washing, cleaning and other activities in order to help the common people in home and also to help the ill people in case of medical emergency [10].

vi. Space: In space technology, robots can be used to operate the planes when the pilots are in need of medical emergency and also to manage the critical situations. Also robots can be used in space exploration and data collection processes [11].

vii. Industrial robots: These robots can be used to monitor the functioning of machines in industry even in case of unavailability of workers. Also automation of various tasks is done with the help of these robots. Effective space utilization can be performed in industries with the help of industrial robots [12].

viii. Military robots: Military robots can be used to replace human warriors during the battle time and also during rescue and crisis periods. These robots can be programmed to fight with enemies or can work autonomously in order to save the life of soldiers [13].

ix. Scalable robots: These robots can work independently and also take decisions by their own in order to solve the problem. These robots first analyze the environment and then use their knowledge to solve the problem of any type [14].

The combination of AI and Robotics has drifted from basic reasoning toward human centric cognitive behavior in order to enhance productivity. In Robotics, perception is one of the most significant applications of AI algorithms where the robots act intellectually by sensing the environment with the assistance of embedded sensors or

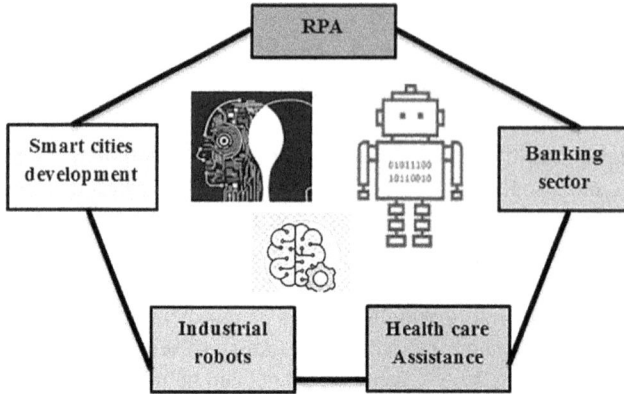

FIGURE 4.3   Applications of AI in Robotics.

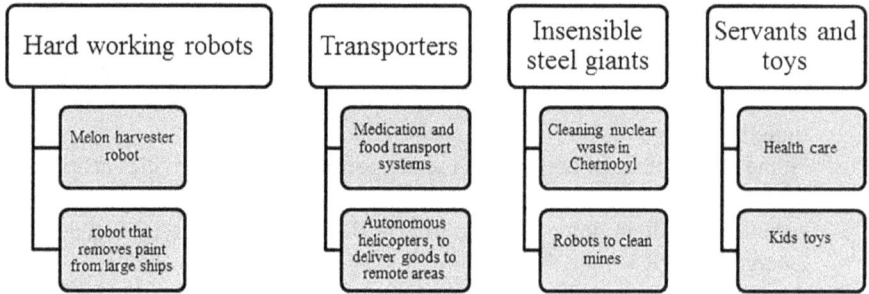

FIGURE 4.4   Types of robots.

computer vision. Such systems have been improving the standard in terms of both sensing and perception. The current section outlines the applications of AI in Robotics (see Figure 4.3) in terms of robotic process automation (RPA) in varied sectors.

Traditional robots can be transformed into advanced robots in four different categories: hard working robots, transporters, insensible steel giants, servants and toys, each of which can be used in several applications. Hard working robots are designed to computerize assembly line work in automobile and agriculture sectors. Transporters are programmed to work based on the regular landmarks. Steels giants are designed exclusively to handle hazardous environment. Servants and toys can be utilized for simple smart home automations and in kid's toys. The applications of the modern robots are summarized in this section (see Figure 4.4) [15].

4.3   ROBOTIC PROCESS AUTOMATION (RPA)

The integration of two phenomenal technologies, namely AI and Robotics, has helped industries to upgrade the existing automation systems in order to handle more intricate and high-level applications. Though the idea of physical robots had come

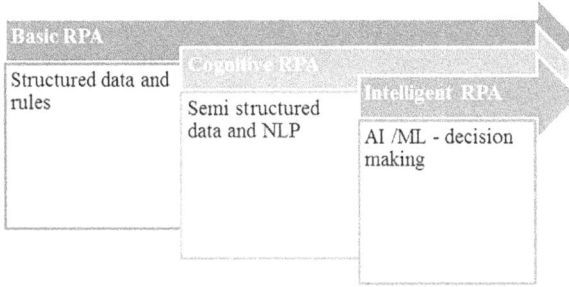

FIGURE 4.5   Revolution of RPA.

into existence in the earlier years, without the functionalities of AI it is not possible to handle such complex automations. The main objective of inclusion of AI in Robotics is making the robots to think and act on their own. This integration has paved the way for the applications like mining techniques, computer vision, Natural Language Processing (NLP), motion control and navigation systems.

The transformation of RPA from the basic model to advanced AI-based RPA is clearly shown in Figure 4.5 (see Figure 4.5). The traditional RPA system deals with the automation of iterative actions by following the same set of instructions without altering the plan of existing structures. The emergence of cognitive RPA paved the way for enormous NLP applications that deal with semi-structured data. An intelligent RPA system incorporates both Machine Learning and Deep Learning algorithms in order to make the robots make conclusions like humans and derive insights for real-time use cases. Therefore, it is apparent that AI technology is disrupting by inheriting the cognitive experiences in rule-based RPA technologies. Thus incorporating, AI in Robotics in terms of augmented intelligence, assisted intelligence and autonomous intelligence.

4.3.1   RPA IN BANKING SECTOR

The two major technologies play an important role in endowing the banks to promptly industrialize their business processes and effectually renovate the complete business. AI-enabled RPA systems provide numerous services in the form of managing customer data, transaction operations, trading, risk and security management. The main objective of incorporating RPA and AI in banking is to lessen the monotonous work by humans and thereby letting people to ponder on difficult and innovative tasks which machines fail to deal with, for example – retaining the customer affiliations. Latent use cases AI and RPA in banking sector includes chat bots for improving customer and banking services, recommendation systems, human traders replaced with AI-based traders and robot advisers. The future of banking services is most expected to be absolutely automated, with the presence of human involvement only involved in terms of critical decision making assisted by perceptions and recommendations from Robotics and AI technologies [16].

4.3.2 AI and Robotics in Healthcare

Smart home systems incorporate the applications of embedded sensors, automations, activity recognition, object recognition gesture recognition and intelligent decision-making systems. Smart homes are basically equipped with sensors and actuators that monitor the occupant's physical activities and the surroundings in terms of home automation and health monitoring [17]. Such systems look after the common activities of daily living and provide an independent life for elderly with the assistance of smart homes [18].

Pharmaceutical companies utilize AI systems to achieve a range of medical processes like drug discovery, drug analysis, disease diagnosis, laboratory results analysis and surgery-oriented analysis. Technology giants like IBM Watson have proven to be more precise than human doctors in their responsibilities. Similarly Japan's Robot Robear can support patients and elderly in basic activities like getting out of bed, walking and many [19].

4.3.3 AI-Based Industrial Robots

The industrial division is exposed to the fourth industrial revolution (Industry 4.0) in which AI-related capabilities are holding an enormous part for the improvement of Industry 4.0 that demands the applications of robotic systems embedded with the combination AI and ML algorithms. The substantial role of utilizing intellectual machines exists in the need of making a smart industry that diminish the lead time and aims at higher productivity with enhanced product quality. The simple industrial use cases that are taken care by industrial robots are manufacturing, data collection, warehousing and integration of people and product services, logistics and delivery services [20].

AI algorithms improve the learning experience by storing the set of repeated tasks so that the robots can teach themselves in the way how to respond and perform a task within a short duration. One of the main reasons for incorporating AI in industrial Robotics is the usefulness of training the machines, whereas the conventional procedure would involve teaching the automation system by means of several rules to differentiate different parts to pick up; this demands implementing much iteration and undergoing numerous trial and error methods. It is evident that the traditional methods are human dependent and time-consuming process; hence AI and ML algorithms can make the design of industrial robots easier to train, and thereby industrial productivity can also be increased [20].

Industrial robots exist in five different categories, each of which varies in processing speed, capacity and workspace. Knowledge and working of each operating characteristic of all five categories (see Figure 4.6) can aid industrial engineers to identify the suitable robot based on their needs and demand [19].

Cartesian robots are the most commonly used robots in industrial sectors, particularly for plant operators. These robots create the workspace in cube-shaped structures where applications are placed within the range of 100 millimeters to 10 meters. Cartesian robots can be easily customized based on the user requirement. Cylindrical robots are the simplest of all the categories that comprise of two different actuators.

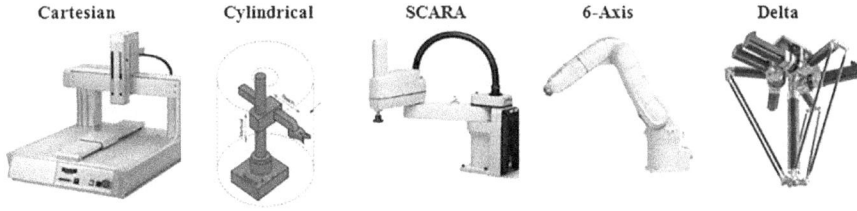

FIGURE 4.6 Types of industrial robots.

The linear actuator takes care of linear category of motions, whereas rotary actuator takes care of rotation movements that enable the robot to move from one place to another. SCARA robots are more advanced than Cartesian and cylindrical robots, as they are armed with three dimensional axes in addition to the rotatory actuators. Yet Cartesian robots are found to the commonly used industrial robots due to their custom features. Six-axis robots are large, toy-sized robots that can act like a human arm and move things or materials from one place to another. Such robots are widely used in automobile sectors, such as fitting of engine parts in cars, fitting the seats on the cars and many other applications [19]. Delta robots are also capable of shifting a group of things from one place to another, used particularly in pharmaceutical and packaging industry. They are considered to be the high-speed robots that are able to differentiate with the help of vision technologies; these robots are able to choose diverse size, shape and color choices and follow a programmed pattern to pick and place things. These robots are capable of replacing the long time labor-based procedures where around hundreds of parts per minute are being picked, organized and positioned. Delta robots are utilized not only to expand reliability and excellence of procedures where monotonous movements eventually lead to employee exhaustion, but they also stand as a solution for several health issues like skeletal disorders, motion injuries, strains related to shoulders, elbow and lower back [21]. These are the five robots that are commonly utilized in industrial sectors to offer smart solutions and to solve numerous issues and save labor time without compromising the quality and reliability of work.

The integration of AI and Robotics has led to the development of new type of learning systems, probably known as robotic learning. There are five significant areas (see Figure 4.7) in Robotics where Machine Learning has had a substantial influence on industrial Robotics. Computer vision allows machines to identify and organize objects based on different properties like movements, shape, size and color. Anomaly detection with neural networks and convolution neural networks is an excellent example for computer vision technologies. Imitation learning, the category of reinforcement leaning, aims to mimic human behavior, particularly toddlers and infants, in solving a given task. The machine is trained to execute a task from demonstrations by learning a plotting amid observations and actions that make the robot to act in the suitable environment by improving the rewards. Self-supervised learning methods allow robots to engender their own training samples to increase the testing results in terms of performance, using a pre-trained test train data taken close to infer

Computer vision	Imitation learning	Self-supervised learning	Assistive technologies	Multi-agent learning

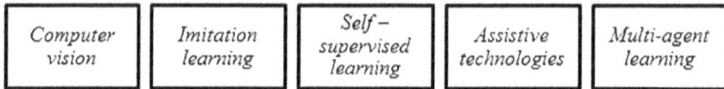

FIGURE 4.7 Significant areas of ML in robotic learning.

long-range abstruse sensor data where it is combined into robots and ophthalmic devices to discover items. Autonomous vehicles that involve the use of unsupervised and reinforcement learning algorithms are considered to be the variant of self-supervised learning. Assistive technologies are exactly suitable for healthcare sectors in numerous use cases like disease diagnosis and elderly healthcare. Multi-agent learning is another class of reinforcement learning that allows several agents to interact in a shared environment. The multiple agents within the common environment are allowed to interact with one another and execute tasks by collectively learning, observing and co-coordinating the multiple outcomes [20].

4.3.4 AI AND ROBOTICS IN SMART CITIES DEVELOPMENT

Smart city policies urbanized by cities around the world deliver a suitable resource for intuitions into the future of smart improvement. The smart city conception is non-static, progresses over time that encompasses human and social capital, durable infrastructure and ICT knowledge in the provision of an ecological environment. The five main applications of AI and Robotics in major countries like Australia, Canada, China, Dubai and UK are decision making, automation, education, smart infrastructure and mobility. With the help of AI and Robotics, certain urban challenges like overcrowded transport, ecological balance, elderly care, security and policing are resolved in smart cities. In terms of smart cities management the following use cases were identified to overcome the communal challenges – Autonomous vehicles allow more effective utilization of transport setup and can deeply diminish the plea for parking in essential zones and clear up the space for housing and reformation. AI-enabled Robotics empowers infrastructures and buildings to react to natural environmental changes, healthcare for elderly and disabled in terms of aided living, Robotics-enabled effective observation, renovation and control, climate control desirable to cope improvements in well-ordered environments for food growing and leisure [22].

4.3.5 DISTINCTION BETWEEN AI AND ROBOTICS

Since Robotics and AI are all related technologies, there must be a clear difference between these two domains. Robotics is aimed at development of systems that can work without the intervention of humans whereas AI involves analysis and training of computer systems. The difference between the two terms is quite unclear as Robotics involves the application of AI algorithms to program the robots and vice versa. The differences are listed in Table 4.2.

TABLE 4.2
AI vs. Robotics

S.No.	Aspects/ Characteristics	Artificial Intelligence	Robotics
1.	Data and Processes	Mainly depends on the unstructured data and the processes are circulating around text, pattern, voice, etc.	Robotics also uses unstructured data
2.	System Type	Machine Learning based	Rule based
3.	Type of Business Processes	User centric process	Repetitive and predictable process
4.	Working Style	Repeats the human actions through consistent learning from datasets	Repeats the human actions through some predefined rules
5.	Accuracy	Not fully accurate	Highly accurate
6.	Usage	• Apple's Siri • Netflix • Google's deep mind	• Medical surgical applications • Space and earth exploration • Laboratory research
7.	Relationship	Acts as a bridge between human intelligence and Machine Learning	To improve their functions, AI is used and it is related to the AI field
8.	Purpose	To create a technology that makes the computer to think like a humans	To implement automation by reducing the human intervention
9.	Pros	• Less room for errors • Improved efficiency • Take risks instead of humans Digital Assistance • Faster decision	• Safe load carrying • Good efficiency • High accuracy • Less time consuming
10.	Cons	• High development cost • Unemployment • Makes humans dependent • Lack of out of the box thinking	• High cost • Requires specialized people for maintenance • Leads to unemployment • Damaged or break easily

4.4 MACHINE LEARNING ALGORITHMS FOR ROBOTIC APPLICATIONS

Modern advancement in the field of AI has paved the way for the development of intelligent robots that are more accurate in performing the selection and placement of objects in industries, in making the self-paced drones and the more allied Industrial Internet of Things. In workplace, Machine Learning has been adopted particularly when robotic systems are involved in the business processes [23]. The key reason for the widespread adoption of Machine Learning techniques in robotic systems is mainly due to the successful contributions in developing smarter robots to perform operations such as the assembly of the manufacturing units, select and place operations of the units and control of drone systems. In the subsequent section, an exhaustive description of the research and development carried out using the Machine Learning algorithms in the field of Robotics is given.

4.4.1 MACHINE LEARNING IN BUSINESS PROCESSES

Many organizations have turned from the traditional rule-based programming techniques, as they no longer aid to provide real-time solutions with the complex data. There is a quench to develop smarter systems with increased abilities to solve the huge volumes of intricate data [23]. Various other computational algorithms have been discussed in [24–29]. A quick look at the robotic applications that use Machine Learning algorithms is provided in this section (see Table 4.3).

4.4.2 ROBOTIC USE CASES WITH MACHINE LEARNING

4.4.2.1 Indoor Navigations

The robot developed by a team of researchers in San Francisco resulted in an interesting tool named SKIL Somatic Tool. These robots are trained to choose the

TABLE 4.3
ML-Based Robotic Applications

Product	Algorithm	Organization	Use
Artificial Intelligence way outs	AI Agent AiCoRE and iRSP	AIBrain	Development of smart phones
Cozmo Consumer Robot	Machine Learning algorithms	Anki	To develop emotional engines
Siri	Machine Learning and Deep Learning techniques	Apple	Voice assistant
AI bots	Facebook Artificial Intelligence Research	Facebook	To train the bots with the ability of conciliation
Robots and self-driving cars	GPU-enabled AI with Deep Learning	H2O	Develop autonomous systems
Data and imagery analysis	Machine Learning	Hummingbird Technologies	To implement crop science applications
Commercial robots, Watson IoT system, Q.bo One	Machine Learning	IBM	Service applications such as hospitality, personalized healthcare, finance services
Self-learning chip	Artificial Intelligence	Intel	To develop efficient highly performing robots
Intelligent warehouse robots	Reinforcement learning, Machine Learning	Kindred	To make accurate decisions along with humans in supply chain environments
Joint intuitive robot programming team	Imitative Deep Learning	KUKA	To work in manufacturing units
Smart Robots	Artificial Intelligence	Vicarious	To develop probabilistic generative models

appropriate button on the elevator when it is alerted to do so. A camera has been mounted on the robot to aid with the visionary challenges. Machine Learning techniques are employed to educate the robot about the control panel on the elevator and thus nurture it to interpret the correct floor. This intelligent tool makes use of Long Short-Term Memory of the Recurrent Neural Networks and Convolutional Neural Networks for the fusion of sensors and use in vision systems, respectively. A framework based on reinforcement learning known as RL4J which is based on a guidance system is also an interesting feature of the SKIL Somatic Tool. This tool can be considered as the operating system of the robot in which it is installed and the library RL4J acts as the important engine to drive the predictions made by the robot. SKIL Somatic Tool is a tool that can be used as a complete package to mimic functionalities similar to human beings in robots. The key feature of this package is derived from the fact that it is combined with the traditionally used framework for using robot software, namely Robot Operating System. These kinds of robots are of much interest in today's business processes like indoor navigation to fetch items without the intervention of the human beings [30].

4.4.2.2 Manufacturing Environment

Robots are employed in manufacturing environments, mainly to estimate crashes in the machines used in the factory. In these types of applications, Machine Learning algorithms can be applied to the data collected from the sensors placed on the machineries available in the industries and the systems integrated with Robotics. The sensor data is analyzed in order to interpret the patterns and behavior of the data to find any anomalous nature in the data. Unknown crashes that happen in the manufacturing environments at unexpected schedules may lead to heavy loss in the industry. Thus the solution is to estimate the occurrence of failures beforehand along with the causes. This can be achieved with the help of pattern recognition which can associate the uncharacteristic behavior in the data. When this uncharacteristic behavior of a particular sensor is identified then that piece of the device can be removed and replaced to solve the issue. Conventional methods of statistical modeling have been outdated for providing real-time awareness about failure and thus Machine Learning has taken over the traditional methods and succeeded in providing real-time alerts. These real-time alerts are provided in a considerably good frequency of time using Machine Learning by taking advantage of the cloud platforms. Lower costs incurred in integration of Machine Learning and cloud platforms can be of high use in data analysis and to generate real-time signals. The challenges that industries face with the use of Machine Learning or Deep Learning algorithms are mainly due to the selection of appropriate data cleaning (preprocessing) techniques, proper calibration of the hyperparameters and appropriate selection of right layers or trees according to the technique adopted. The ultimate aim here is to develop machines that can make decisions autonomously with fewer inputs from the humans.

4.4.2.3 Unmanned Aerial Vehicles

The working environment involving drones and unmanned aerial vehicles is comparatively complicated and dangerous in nature. Machine Learning has been adopted wisely to handle such hazardous environments to carry out the tasks with ease.

In order to derive better practices in agricultural fields, Machine Learning with structured approach can be used to extract data from complex images. Drone is used as an important tool to gather field information and this data is then applied to Machine Learning algorithms to arrive at useful insights. The futuristic approach in this area is to embed the machine intelligence in the drone itself to gather and analyze the data on the spot without the need for analyzing it at a later time. In agricultural sector, drones in combination with Machine Learning are used to provide the expected levels of moisture in the soil in order to upgrade the irrigation systems and thus help in preserving gallons of water. Moisture prediction using Machine Learning is tedious as there are several environmental factors that need to be considered along with the weather conditions combined with the data provided by the sensor data from drones.

4.4.3   MODERN ARENAS OF ARTIFICIAL INTELLIGENCE IN ROBOTICS

AI is a vast domain which comprises fields which are sub-specialties used for application development. Some of the modern areas used in robotic applications alongside AI are discussed in this section (see Figure 4.8).

4.4.3.1   Statistical Learning

Statistical learning can be thought of as a tool used in Machine Learning for functional analysis. This is an interesting arena of AI which can be employed in Robotics with the notion of statistical perspective. Bayesian modeling is the most

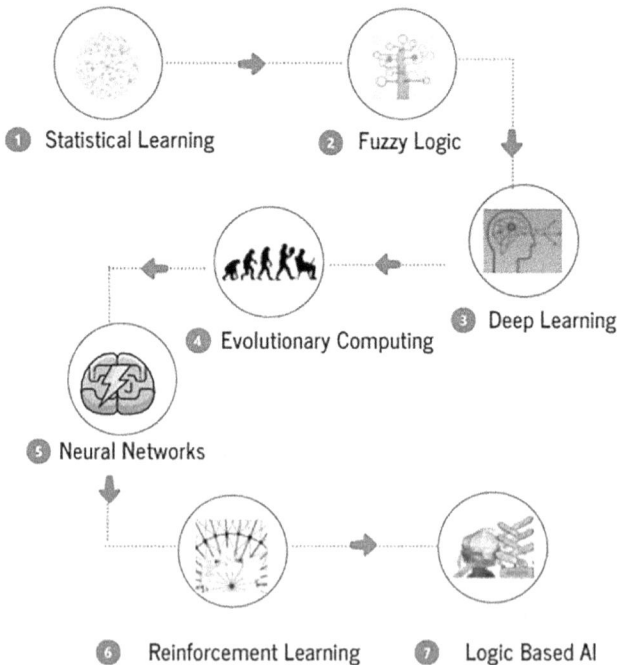

FIGURE 4.8   Modern arenas in Artificial Intelligence.

popular statistical technique which works with the prior knowledge of the classical statistical methods and also helps to develop newer methods which can handle the AI applications. One of the setbacks in this technique is the use of probabilistic methods. These methods cannot prove to obtain the expected results all the time as it is taken as an inference for the whole population. In robotic applications, probability density function is used primarily by the robot to know about the state in which it currently resides, and this knowledge is periodically updated based on the new inputs received from the sensors and from the model created from the results of the prior actions.

4.4.3.2 Fuzzy Logic

This is a kind of reasoning technique which can be used to mimic the reasoning nature of humans in machines. Any decision-making process involves conditions which return two possible results such as either yes or no; similar idea is adopted in Fuzzy logic to imbibe the reasoning capability. Input to the system can be in disordered form which could be incomplete or contain inaccurate data also, but the output produced by the system will be definite. This fuzzy logic method is flexible enough to be implemented in small-scale systems as well as in large control systems. As far as its application in commercial applications is concerned, it can be used to control the operations performed in a machine. The only drawback of this technique is that it does not provide learning ability to the machine for which it should be combined with the other learning methods such as statistical learning or neural networks.

4.4.3.3 Deep Learning

Deep Learning can be imagined to be a subfield of Machine Learning which combines the abilities of neural networks. The learning process in neural networks happens through multiple layers and the data in different varied representations is expected to pass through these multilayers. The architectures in Deep Learning can be used for various applications including computer vision, signal recognition and speech recognition. The various layers in the network include a layer for input and output with a few layers hidden in between the input and the output layer. Each neuron in the network makes use of an activation function in order to normalize the output from the network. Deep Learning is empowered to work on large datasets due to the high computational power. The importance of the input value fed to the network can be determined through the weights assigned to each of these input values. These weights are then adjusted in subsequent iterations to improve the accuracy of the output. Deep Learning is a good choice for robotic applications because of its ability to create non-linear models that can analyze the data, learn its feature attributes and make predictions readily.

4.4.3.4 Evolutionary Computing

Evolutionary computing is adopted in robotic applications to develop robots that are capable enough to adapt to the environment through evolutionary computation process which is close to the process of evolution naturally. Robots are generated by following the principles of evolution and the basic principle pursued for the evolution which is the survival of the fittest. This method is primarily used to design the robotic control systems. In evolutionary Robotics, like other evolutionary techniques, the process begins with a set of population. Each candidate in the population will contain the details of the robot's body plan. There are basically two types of evolution involved, namely, evolution based on novelty and evolution based on fitness, there exists a third variety as well which is a combination of fitness and novelty-based evolution.

4.4.3.5   Neural Networks

Path planning is one of the important problems faced by the robots. Robots are designed with several sensors for vision like camera, and sound recognition is aided by ultrasonic sensors. The data extracted from these sensors are not high-level information or not semantic in nature. With sensors, robots can determine an object which lies a few feet away from them, but it is highly difficult for the robot to make decisions on how to react without knowing the type of object. Neural networks can be very useful for the extraction of high-order information from the low-order non-semantic data. In robot navigation, robot is expected to follow a path to the destination by avoiding the objects that are on the way and reach the goal successfully. Using neural networks for path planning involves three major steps. First, the robot should be fed with a complete knowledge of the environment. Second, it should be capable enough to identify the movements required from the source location to the destination location inside the environment. But it is not possible to identify the entire movable and static objects in the environment beforehand. The real challenge occurs in the last step which can be solved by the planning methods that can be implemented in the local environment.

4.4.3.6   Reinforcement Learning

Robots generally used in the industries are well programmed with careful control systems, but it becomes quite difficult to design robots for domestic needs that can get accustomed to the environment with ease. This can be achieved by the learning algorithm like reinforcement learning that can make the robots comfortable with the unstructured environments. In reinforcement learning, the robot is made to learn through trial and error method to adapt with narrative behaviors. This reduces the burden on the side of the human professionals employed for the operation of the robot. Robot when introduced to the real world only gets more chances of learning from the tasks it performs in the real environment with fewer inputs from humans. Markov Decision Process is a framework commonly used for the learning problems which fall under the reinforcement learning category. On comparing the framework used for robot training and real world environment, there are a few pitfalls in the MDP framework. Reset mechanisms are available in the framework in which the robot can move ahead to different states and start again from the beginning whenever required, but this is not the case in the natural environment. Secondly, there is a huge difference in the dimensionality of the sensors used in the framework and the real world environment. The reward functions are predefined or external in case of the frameworks, whereas the rewards are allocated from their own input sensors in the real case. Thus, it becomes essential to develop robots which can handle all these three challenges in order to be effective and scalable in the real environments.

4.4.3.7   Logic-Based AI

This is a subfield of AI which is mainly used for the representation as well as the inferences made from the knowledge base. Logic programs are used to describe the predicates in the knowledge, the facts represented and the semantics involved in the notion of formal logic. Inductive logic programming is a method employed to derive useful hypotheses from the already available knowledge.

4.5   SOCIETAL IMPACT OF ARTIFICIAL INTELLIGENCE

When it comes to the life part, there are always the positive and negative impacts. AI is ready to convert the whole world as we live in. The balancing factor is the prime question for someone to guess and it might be the debate case for many people. The upcoming modifications will be good in most cases and on the other hand these will be challenging for some people.

4.5.1   AI: POSITIVE IMPACTS

The efficiency of the workplace can be improved with the help of AI. Whenever AI takes on dangerous and repetitive tasks, it frees up their time and allows the humans to think of some creative solutions, and it also increases job satisfaction and happiness among the human community.

There has been much improvement in health-monitoring of humans because of AI. The impact of taking care of the patients has literally reduced the pharma sector to a big margin. Autonomous transportation and AI has influenced traffic congestion and will gain uncountable hours of productivity. This will free up stress among the humans and also help to spend the time for some other productive ways.

Facial recognition is an awesome technology which is just becoming as common as our fingerprints. It solves criminal activity with proper justice and it reveals how good it has been to effectively use the technology without crossing the individual people's privacy [31].

If anyone didn't want to roll out their life with the new inventions then their life will be hugely impacted with AI.

4.5.2   AI: NEGATIVE IMPACTS

4.5.2.1   Change in Human Experience

If AI reduces the human working time by doing many of the tasks full of automation, then the humans should sustain their own social and mental advantages by involving in the new activities [32].

4.5.2.2   Stimulated Hacking

AI exceeds the human ability in many cases in terms of increasing the speed that is too tough to be followed by humans.

4.5.2.3   Loss of Jobs

The change of work nature forces humans to get more training programs since many jobs will be created by AI.

4.5.2.4   Laws and Regulations

The actions taken from one country will affect the other country easily since every country has its own laws and regulations which might not suit others. AI requires updated laws and regulations among different governments.

4.5.2.5   Bias

If there is a bias in training data or in the algorithm, then the result is also biased. AI's must not follow biased logic at any cause since it will affect one set of humans.

4.5.2.6   Terrorism

The new AI technologies like autonomous drones and robotic swarms may create a potential threat among the society if these technologies are not properly used (see Figure 4.9)

4.5.3   Challenges of AI for Robotics

Though AI and its subfields lead to a lot of benefits in robotic applications, there still exist few disadvantages that need to be sorted out. The most important challenges are highlighted in this section.

4.5.3.1   Challenge 1: Volume of Training Data

It becomes immensely difficult to generate training data for the physical processes that occur in the robotic environment. Despite being expensive, this process also consumes a considerable amount of time. One of the solutions to this problem is to generate the data used for training purpose through simulation technique [33]. Another solution includes the incorporation of digital manipulation in order to leverage the training data.

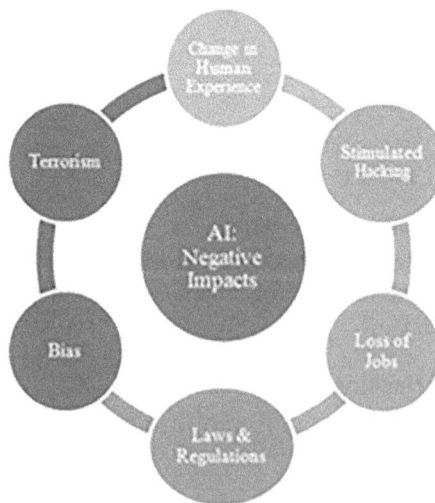

FIGURE 4.9   AI negative impacts.

4.5.3.2   Challenge 2: Time Taken to Train

This is yet another challenge that is associated with the use of AI for robotic applications. The training process usually involves several millions of parameters and thus takes even days to complete the training process. One approach that can be adopted to reduce the training time includes the distribution of tasks among multiple units. This kind of an approach will be very useful to improve the speed of training as well the classification process. Another solution to this problem includes the usage of two-step approach where few parameters are employed in one step followed by increased parameters in the second step, thus improving the performance of the system.

4.5.3.3   Challenge 3: Estimating Uncertainty

In order to employ Deep Learning for robotic applications, it becomes highly essential to approximate the uncertainty in the decisions made by the learner. Estimating uncertainty is important in a way that only then the system will consider the neural network similar to how it considers the other sensors and employ the algorithms to make accurate predictions.

4.5.3.4   Challenge 4: Unknown's Identification

A general assumption when any Deep Learning algorithm is employed is that the model which is trained for certain conditions will be deployed for the same set of conditions for which it was trained earlier. But this is not the case with robots because the environment in which it is deployed may vary and is uncontrollable in real time in nature [34]. So, there is a high possibility to come across scenarios or conditions that were not encountered during the training phase. Thus, it becomes difficult to recognize unknown objects and there are chances for the learner to assign high confidence scores for false objects.

4.5.3.5   Challenge 5: Continuous Learning

In many robotic applications, the objects in the real world environment may deviate in characteristics as well as appearance compared to the exposure it had during the training phase [34]. This problem can be rectified by making the system in robotic vision kind of applications to adapt itself to the new training examples of the already known target values and learn its representations instantly.

Despite these challenges, AI technology is getting updated on a regular basis which can pave the way to encounter these issues with ease. Artificial General Intelligence (AGI) is considered to be a kind of hypothetical intelligence which involves leveraging human intellectuals to the machine and thus making it to behave and act like humans. In this technology, the robot has to interact with the environment in the right way to make correct decisions. The main advantage of combining AI with Robotics lies in the fact that in this way we make the systems learn from their own errors, and system automatically corrects the errors as it proceeds with the work. Minor issues can be dealt by the system itself and only drastic problems require the involvement of the human supervisor. Monitoring the activities of the robot is much easier for humans than to perform those activities manually.

4.6 CONCLUSION

The increased competencies and civilization of AI systems have paved the way for the development of tremendous applications in diverse sectors. The traditional AI algorithms like fuzzy logics, genetic algorithms, neural networks, probability and statistics have transformed to subfields like Machine Learning, Deep Learning, reinforcement learning, ensemble learning, meta algorithms and evolutionary computing techniques that have created a new era for technological interventions. Robotics is one such sector that utilizes AI in the form of specialized AI, especially for industrial robots to execute repetitive jobs. It is highly evident that AI-based robotic solutions can highly diminish the error rate, without compromising the efficiency and accuracy. The integration of AI and Robotics leads to the development of efficient and profitable automation applications in robotic processes based on four key processes – vision, grasping and motion control. The foremost use cases of AI-based Robotics, particularly in industrial sectors, are logistics and supply chain management. The other sectors also benefit from these integrated technologies in diverse use cases like motion control, computer vision, navigation systems and NLP.

The impact of AI in Robotics involves the entire life transformation, yet there are certain drawbacks to be addressed here. AI-based robotic applications are highly cost effective; nevertheless, the technology demands enormous early investments owing to its complexity. Also the technological transformations can change the livelihood of baseline workers, as they will be replaced with machines, excluding the positions that require creativity. Yet robots can save huge time and efforts, handle complex tasks and offer efficient outcomes. Based on the recent survey, it is predicted that by 2025, the AI–Robotics combo will be incorporated in every segment of our lives, also robots will be commercialized and trained to handle customers in business, marketing, hazardous tasks in industries, and they will be sharing a common workplace. This is how AI and Machine Learning are fixed to transmute the Robotics sector in an enormous manner.

REFERENCES

[1] Turing, A.M. "Computing Machinery and Intelligence." In Epstein, R., Roberts, G., Beber, G. (eds) *Parsing the Turing Test* (2009). Springer, Dordrecht. https://doi.org/10.1007/978-1-4020-6710-5_3.
[2] Yang, L. "Research on Application of Artificial Intelligence Based on Big Data Background in Computer Network Technology." *IOP Conference Series: Materials Science and Engineering*, 2018, 392, 062185.
[3] https://www.techopedia.com/definition/31621/weak-Artificial-intelligence-weak-ai.
[4] https://www.ibm.com/cloud/learn/strong-ai.
[5] https://www.techopedia.com/definition/31619/Artificial-superintelligence-asi.
[6] Mohammad, S. M."AI Automation and Application in Diverse Sectors." *International Journal of Computer Trends and Technology (IJCTT)*, 2020, 68(1).
[7] Sood, M., Leichtle, S. W. *Essentials of Robotic Surgery*. Spry Publishing LLC, 2013.
[8] Halundi, A. M., et al. "The Robotics and Applications." *International Journal of Computer Science Engineering (IJCSE)*, July–August 2018, 7(4), 150–153.

[9] Joseph, L., Joseph, S., Sarath, H., Abraham, S.P., George, A. "An Autonomous Agriculture Robot Using IoT." *International Journal of Computer Sciences and Engineering*, 2020, 8(7), 49–53.

[10] Gerhart, J. *Home Automation and Wiring*. McGraw Hill Professional, 1999.

[11] Launius, R. D., McCurdy, H. E. *Robots in Space: Technology, Evolution, and Interplanetary Travel*. The Johns Hopkins University Press, Baltimore, 2008.

[12] Nof, S. Y. *Springer Handbook of Automation*. Springer Verlag, 2008.

[13] Springer, P. J. *Military Robots and Drones: A Reference Handbook*. ABC-CLIO Editor, 2013.

[14] Stay, K., Brandt, D., Christensen, D. J. *Self-Reconfigurable Robots: An Introduction*. MIT Press, 2010.

[15] Sahare, P.H., Kumbhalkar, M.A., Nandgaye, D.B., Mate, S.V., Nasare, H.A., "Concept of Artificial Intelligence in Various Application of Robotics." In *International Conference on Management and Artificial Intelligence IPEDR* (vol. 6, 2011). IACSIT Press, Bali, Indonesia.

[16] https://ibsintelligence.com/files/media_temp/HexawareRole_of_AI&RPA_in_trans-forming_Banking_Operations_9.pdf.

[17] Bennett, J., Rokas, O., Chen, L. "Healthcare in the Smart Home: A Study of Past, Present and Future." *Sustainability*, 2017, 9, 840.

[18] Tian, S., Yang, W., Le Grange, J. M., Wang, P., Huang, W., Ye, Z. "Smart Healthcare: Making Medical Care More Intelligent." *Global Health Journal*, September 2019, 3(3), 62–65.

[19] Dubey, D., Dewangan, U. K., Soni, M., Narang, M. K. "An Investigation of Application of Artificial Intelligence in Robotics." *International Research Journal of Engineering and Technology (IRJET)*, July 2019, 6(7), 1571–1577.

[20] Benotsmane, R., László, D., György, K. "Survey on Artificial Intelligence Algorithms Used in Industrial Robotics." *Multidiszciplináris Tudományok*, 2020, 10(4), 194–205.

[21] https://motioncontrolsRobotics.com/delta-robots/.

[22] Golubchikov, O., Thornbush, M. "Artificial Intelligence and Robotics in Smart City Strategies and Planned Smart Development." *Smart Cities*, 2020, 3(4), 1133–1144.

[23] Karthiga, M., Sountharrajan, S., Nandhini, S. S., Kumar, B. S. "Machine Learning Based Diagnosis of Alzheimer's Disease." In *International Conference on Image Processing and Capsule Networks* (pp. 607–619, May 2020). Springer, Cham.

[24] Sountharrajan, S., Suganya, E., Karthiga, M., Nandhini, S. S., Vishnupriya, B., Sathiskumar, B. "On-the-Go Network Establishment of IoT Devices to Meet the Need of Processing Big Data Using Machine Learning Algorithms." In *Business Intelligence for Enterprise Internet of Things* (pp. 151–168, 2020). Springer, Cham.

[25] Karthiga, M., Sountharrajan, S., Nandhini, S. S., Suganya, E., Sankarananth, S. "A Deep Learning Approach to Classify the Honeybee Species and Health Identification." In *2021 Seventh International Conference on Bio Signals, Images, and Instrumentation (ICBSII)* (pp. 1–7, March 2021). IEEE.

[26] Karthiga, M., Nandhini, S. S., Tharsanee, R. M., Nivaashini, M., Soundariya, R. S. "Blockchain for Automotive Security and Privacy with Related Use Cases." In *Transforming Cybersecurity Solutions Using Blockchain* (pp. 185–214, 2021). Springer, Singapore.

[27] Karthiga, M., Sankarananth, S., Sountharrajan, S., Kumar, B. S., Nandhini, S. S. "Challenges and Opportunities of Big Data Integration in Patient-Centric Healthcare Analytics Using Mobile Networks." In *Demystifying Big Data, Machine Learning, and Deep Learning for Healthcare Analytics* (pp. 85–108, 2021). Elsevier.

[28] Suganya, E., Sountharrajan, S., Shandilya, S. K., Karthiga, M. "IoT in Agriculture Investigation on Plant Diseases and Nutrient Level Using Image Analysis Techniques." In *Internet of Things in Biomedical Engineering* (pp. 117–130, 2019). Springer, Cham.

[29] Sountharrajan, S., Nivashini, M., Shandilya, S. K., Suganya, E., Banu, A. B., Karthiga, M. "Dynamic Recognition of Phishing URLS Using Deep Learning Techniques." In *Advances in Cyber Security Analytics and Decision Systems* (pp. 27–56, 2020). Springer, Cham.

[30] https://www.Roboticsbusinessreview.com/wp-content/uploads/2018/04/RBR_MachineLearningRobots_WP_Final.pdf.

[31] https://bernardmarr.com/what-are-the-negative-impacts-of-Artificial-intelligence-ai.

[32] https://www.leewayhertz.com/ai-applications-across-major-industries.

[33] Pierson, H., Gashler, M. "Deep Learning in Robotics: A Review of Recent Research." *Advanced Robotics*, 2017, 31(16), 821–835, https://doi.org/10.1080/01691864.2017.1365009.

[34] Sünderhauf, N., Brock, O., Scheirer, W., et al. "The Limits and Potentials of Deep Learning for Robotics." *The International Journal of Robotics Research*, 2018, 37(4–5), 405–420, https://doi.org/10.1177/0278364918770733.

5 Robotic Process Automation in COVID-19

P. Naveena, T. Pradeepika, V. Priyadharshini, and A. Padmashree

Bannari Amman Institute of Technology, Erode, India

CONTENTS

DOI: 10.1201/9781003181668-5

5.1   INTRODUCTION

The COVID-19 pandemic shocked the whole world. Governments and organizations hurriedly executed measures to ensure general wellbeing just as the nations' economies. Numerous organizations shut down their workplaces and their processing plants. Most middle-class representatives began telecommuting, and online specialized apparatuses became key parts of their lives. While governments and organizations were adjusting to the new common, flexible superior representatives, product robots have been more apparent. Robotic Process Automation (RPA) presented the ultra-performing and quick-learning programming robots into our lives. These robots can work day in and day out in numerous spaces, helping human representatives and aiding clients. This section talks about various uses of RPA robots throughout the planet and in Turkey during the COVID-19 pandemic. RPA turned into the new problem-solving innovation that offered development to the business world. It discovered application regions during COVID-19 in different areas, for example, medical services, instruction and public areas. Governments and organizations found support from the RPA robots to follow patients, advance production network measures, help understudies and instructors in online classes and react to the expanded interest in internet shopping.

5.2   WHAT IS RPA?

RPA is the utilization of programming that computerizes manual errands. These "bots" permit associations to robotize straightforward or redundant cycles to diminish the time spent on exorbitant manual assignments and increment endeavors to convey crucial work. The product is intended to perform routine undertakings across various applications and frameworks inside a current work process. It performs explicit undertakings to mechanize the exchange, altering, detailing and additionally saving information. Where there is paper, manual assignments and complex work process ventures, there is rich freedom to infuse RPA to further develop exactness, abbreviate measures from days to minutes and to drastically further develop business execution. This effectiveness can eventually save organizations basic dollars where it really matters, permitting organizations to scale rapidly to completely fulfill need, make do with diminished headcount or expand speed and precision. RPA bots can work freely or close by people. Bots have taken over work cooperatively with people to further develop speed and exactness of the modest parts of their work, while unattended bots can work self-sufficiently in the absence of any human intervention. For instance, in a client support setting, They've gone to the bots can be utilized to recover client information quicker, permitting every client support representative to work through a long line all the more proficiently, subsequently holding client reliability. As often as possible, bots are sent to automate administrative center tasks, including information gathering, handling, and analysis. Envision, for instance, a clinic framework endeavoring to extend the quantity of ventilators they will require during COVID-19 regardless of whether to divert supply of these important assets starting with one medical clinic then onto the next. An unattended bot could gather information from emergency clinic affirmations and clinical diagrams to assist with deciding the quantity of assets needed for every office.

5.3   WHAT ADVANTAGES DOES RPA BRING TO A COVID-19 ENVIRONMENT?

The pressing factors of ordinary business tasks have just been amplified by the current environment. Numerous organizations have encountered critical income decreases and are feeling the squeeze to check costs and do more with less assets. RPA can help organizations address the accompanying business needs:

Cost investment funds: RPA ventures can create massive expense reserve funds. Return on initial capital investment is acknowledged close for a moment, counterbalancing the forthright venture. A more modest execution with 10 bots or less can be carried out generally cheaply and inside a brief timeframe. RPA can expand the amount and nature of work item, while permitting human resource assets to move to higher-esteem assignments or be redeployed to different pieces of the business—all adding to the financial matters of computerization. In the current environment, this is basic.

Speed: RPA can cut the time spent on manual undertakings by significant degrees—a fundamental need during the pandemic when lost time can mean death toll, just like the case for medical service suppliers and producers of basic clinical supplies.

Usefulness: As incomes have dropped, numerous organizations have been compelled to lay off laborers. Keeping up with usefulness with less assets is an absolute necessity. Via robotizing parts of the labor force's everyday exercises, staff can rather zero in on exercises that require human critical thinking.

Going virtual: With all unimportant organizations compelled to stop face-to-face work and the course of events until an immunization is accessible probable one or two years away, organizations should work with exceptionally powerful distant work quickly. RPA can be utilized to speed up the arrangement interaction, guaranteeing workers approach proper Wi-Fi at home and are enrolled for new gear at work spaces.

Business progression: Sometimes, robotization can secure staff's actual wellbeing by restricting openness. For instance, fundamental organizations are utilizing bots to assess every representative's present wellbeing and COVID-19 danger. In view of the study reactions, it decides every day if the individual is adequately okay to go into work. This utilization of RPA could become supportive when portions of the nation look toward a staged resuming and need to forestall inescapable disease in the work environment.

Exactness: RPA permits you to take out the human wiggle room, which on account of monotonous undertakings is set somewhere in the range of 5%–10%. Further developed exactness and nature of work item is fundamental in high-stakes errands.

5.4   ANALYZING THE PROS AND CONS OF ROBOTIC PROCESS AUTOMATION

At this point, even the least technically knowledgeable among us has known about man-made consciousness, AI and mechanical cycle mechanization (RPA). These sorts of apparatuses drive a specific measure of dread and vulnerability among workers who stress their positions may sometime be supplanted by robots. However, that innovation, explicitly RPA, can hold extraordinary guarantee for the labor force.

5.4.1   PANDEMIC DRIVES A MOVE TOWARD INCREASED AUTOMATION

RPA has expanded in ubiquity as the COVID-19 pandemic focused on the requirement for effectiveness and cycle improvement, said Emily Rose McRae, ranking executive of the Gartner HR practice. She highlighted aftereffects of a 2020 Gartner COVID-19 Quick Poll, in which 24% of senior money pioneers detailed that they intended to build their interests in RPA, work process mechanization and improvement innovations because of the pandemic. Another 68% said they were intending to keep up with their present speculation levels. "As associations enter post-pandemic recuperation and try to resume worksites, arising innovations will keep on assuming a crucial part in supporting better approaches for working," McRae said. Yet, similar to any tech arrangement, RPA is certainly not a silver slug and may not be appropriate for each association. The other side of this is likewise obvious: Every association may not be correct, or prepared, for RPA. Here, we utilize a straightforward SWOT—qualities, shortcomings, openings, dangers—examination to decide how RPA can possibly hold an ideal spot inside your HR activities.

5.4.2 Potential Strengths

The undeniable huge advantage of RPA is that it does authoritative jobs so individuals can zero in on more significant level spaces of effect. "Innovation, like mechanical cycle mechanization, works best when it upgrades human capacity, saving time for individuals to zero in on undertakings that require basic reasoning, judgment and sympathy—instead of straightforward, value-based work," said Manish Sharma, bunch CEOs of Accenture Operations, where he drives a group of 145,000 experts around the world. According to Sharma, Accenture research has discovered that "HR pioneers from the most carefully developed associations perceive that mechanization is tied in with amplifying ability in a period when individuals are generally critical to its prosperity."

5.4.3 Potential Weaknesses

In any case, RPA may not generally give the advantages trusted for. Sharma based his study on "absence of authority sponsorship, siloed conduct and ability holes" as reasons mechanization might flounder. Critically, RPA—in truth, any innovation arrangement—isn't something that just "naturally" works. Since you have an instrument set up. It doesn't mean that it will be used effectively. The HR chiefs at Accenture who were surveyed demonstrated that additional reskilling should be done to guarantee apparatuses are being utilized most fittingly and adequately. Also, not simply reskilling identified with the tech apparatuses. "The end worth of a mechanized activity normally contacts various cycles, so you need between usefulness across divisions," Sharma noted. For instance, he says, "a CFO and a CHRO can't change their capacity if the CIO isn't ready." Overcoming possible shortcomings with RPA, he said, "It's about coordinated effort, cycle, and innovation." There are other inborn shortcomings in certain associations that could affect the capacity to adequately organize RPA. As an example, McRae emphasized "expanding interest in computerized expertise and social-inventive abilities." This is significant as representatives start "working close by computerization instruments and dealing with the yield of RPA-empowered cycles." HR pioneers, she said, ought to consider "the best approaches to improve advanced aptitude and fabricate social-imaginative abilities where RPA is utilized." Another potential shortcoming could be chiefs ill-equipped to track the progress of the RPA . Administrators "will confront new difficulties in estimating representative usefulness, keeping a positive culture and speaking with groups in regards to the advantages and difficulties that might the accompanying mechanization process began," McRae said. It's significant, "she added," for HR to help administrators, not only workers, through these changes.

5.4.4 Potential Opportunities

The pandemic has highlighted one basic chance of RPA for HR associations: the capacity to stay useful notwithstanding staffing deficiencies and a scattered labor force. "Robotization is particularly important in the midst of emergency, for example, the COVID-19 pandemic we are as yet confronting today," Sharma said. "Truth be told, 70% of the HR chiefs we studied say computerization is comprehensively being utilized today." As we rise up out of the pandemic, Sharma said, HR pioneers are becoming planners of "another, mixed, human-in addition to machine labor force." This expects HR to "match the best human ability with the innovation they need to dominate," he said. "Future-prepared HR leaders have realized rapidly that computerization will be an empowering influence of unmatched advances in worker efficiency and commitment," Sharma said. Fred Hencke, senior VP at Segal, a HR and worker benefits counseling firm in New York City, said RPA can emphatically affect cost, quality and versatility when it is centered appropriately, works mistake free and doesn't create countless special cases for be dealt with physically. To best use these likely chances, he suggested first guiding RPA use prior to organizing it comprehensively. "Looking forward, Gartner research uncovers 82% of associations will include heads of HR consequently to-work choices, so it is significant that HR chiefs know about the innovations that will empower the post-pandemic universe of work and lead the association in getting ready for the effect they will have on the labor force," McRae said. That is absolutely a chance. It can likewise end up being a danger, however, on the off chance that other basic strategic policies aren't set up.

5.5 THE MEANING OF RPA IN HEALTHCARE INDUSTRY

Medical services frameworks contain numerous oppressive assignments that require significant measure of asset designation, for example, guarantee the board. This prompts significant expenses of tasks and moderate cycles. Utilizing the force of mechanization and RPA, medical care suppliers can resolve these issues and make medical services frameworks more effective and medical services measures quicker, working on understanding fulfillment.

5.5.1 WHY IS RPA SIGNIFICANT IN MEDICAL CARE?

Since hospital therapy is perhaps the maximum wasteful ventures, decreasing medical services failures will add to better medical care conveyance which is significant for both the business and people in general. Each industry has failures; anyway, a couple of businesses face the difficulties of medical services industry: severe guidelines concerning patient information and less assets to manage such guidelines. Monetary administrations additionally face comparable significant degrees of guidelines; however, they have good admittance to them and had truly had more elevated types of innovation speculation with Goldman Sachs CEO calling the organization as an innovation organization. Accordingly, the degree of failures and manual cycles in medical services is higher than practically some other industry. IT and medical care administrations financial plans all come from medical services suppliers' income. With robotization and quick execution projects empowered by RPA, medical care suppliers can keep away from expensive, long-running computerized change execution projects and receive quick benefits, empowering them to channel more assets to medical care conveyance.

5.5.2   Use Cases of RPA in Medical Care

5.5.2.1   Patient Scheduling

By the inclusion of RPA innovation, the sufferer can plan the arrangements by not taking a medication from medical clinic representatives. Alongside wiping out the need of asset assignment for planning, this application can likewise further develop client relations so that the sufferer can orchestrate an arrangement quicker.

5.5.2.2   Guarantee Administration

As soon as a medical care administration is given, charging sets aside time because of manual and dull assignments in guaranteeing the board cycle. Guaranteeing the board contains cycles, for example, contributing, preparing and assessing archives and information. Alongside robotizing time-concentrated errands, RPA-drove guarantees the executives can likewise take out human blunders during guarantee handling. As indicated by examines, Medicare/Medicaid protection fakes are most of bogus cases among any remaining protection fakes in America.

More than 80% of companies are accelerating automation in responce to COVID-19

Percentage of companies taking action to accelerate automation initiatives

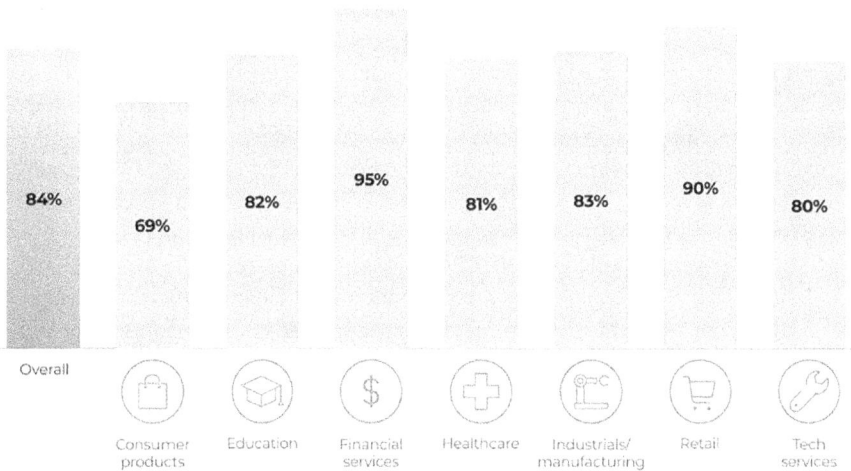

Overall	Consumer products	Education	Financial services	Healthcare	Industrials/ manufacturing	Retail	Tech services
84%	69%	82%	95%	81%	83%	90%	80%

Source: Bain COVID-19 IT buyer survey.

5.5.2.3   Regulatory Compliance

RPA empowers medical care suppliers to observe and file every cycle step in prepared log files so that the corporation can agree to outside reviews. Since these cycles are dealt with by bots, RPA improves information classification too.

5.5.2.4 Resettling/Relocation

"Medical services suppliers" is dependent on the archives and needs advanced change. Medical services suppliers are digitizing patient data with the goal that it tends to be put away electronically and accessed online by different specialists and the actual patients. The way toward removing information from inheritance frameworks and creating that into computerized frameworks could be robotized by robots. Then in a while, while relocation of statistics for another cause like clinical exploration, other robots can deal with those sorts of movement interactions.

5.5.2.5 Use Cases of Skeptic Industry

Cycles similarly owned, cost the board and so forth are normal among all ventures and RPA can be utilized to computerize them. For all the more such RPA use cases, go ahead and check our article where we clarify 60+ RPA use cases.

5.6 WHAT ARE THE BENEFITS OF RPA IN HEALTHCARE?

Top advantages of RPA in medical services industry are as follows:

- Lessening costs: Price of RPA writing computer programs is just a little piece of what clinical consideration providers pay to delegates for manual endeavors. As shown by CAQH's examination, the clinical benefits industry could save $13.3 billion if administrative tasks are robotized in the pay cycle.
- Increased course of action turnout: Thanks to the robotization of patient arranging and sport plan refreshes ship through RPA bot, sufferers are less disposed to overlook to consider their sport plans so experts provide attention to extra patients that grows handiness and adequacy.
- Elimination of human bungle: In rules-based cycles, bots apply conclude that are adjusted. If computer programmer doesn't make any bungle while creating code, your rules-based cycle will be without botch.
- Better patient experience: The RPA bots smooth out front-office support and simplify it for affected person assist bunch with directing patient requests. RPA game plan toward the front and the authoritative focus licenses clinical consideration providers to offer a more incredible customer support.
- Better representative fulfillment: Assigning your labor force to dreary undertakings might hurt worker fulfillment that can prompt higher representative turnover. This prompts seriously enrolling and on boarding during which workers are significantly less useful.

5.7 REAL-WORLD APPLICATION OF RPA IN HEALTHCARE

RPA use cases in clinical consideration offer significant information into why it justifies passing on some dull tasks, for example, entering patients' blood test results into express records and thereafter dependably invigorating those archives to programming robots. Surely, by acknowledging which endeavors are better managed by bots, the "why" is unquestionably answered. Robotized measures are faster, more streamlined and less harmed by bumbles, so customers—the patients, rush to benefit. Subsequently, the real community and its laborers will see the will increase and, totally probably, nonetheless open to impeach to continue with the motorization adventure.

5.7.1 HIT THE HIGH SPOTS OF PATIENTS' BOOKING

Programming robots can smooth out web-based planning. Elements got through the arrangement demand, similar to conclusion, area, protection transporter, individual inclinations and so on, can be accumulated in a report and dispatched for a reference to the board delegate who really makes the arrangement.

Who will profit? Basically, everyone: a simpler occupation for the call place staff, less errors, more fulfilled clients and all the more equitably conveyed arrangements across specialists' functioning time.

5.7.2 MORE VIABLE ADMINISTRATION OF SUPPLY MEASURES (CLAIMS AND CHARGING)

Cases the board takes up a ton of hospital therapy chairmen's time, if via a few stroke of exact good fortune because of the various sub-measures that ought to be looked after: contributing, managing, assessing, managing claims. Computerization can prompt generously quicker and mistake-free preparation, finally putting down chairmen of apparent weight.

Additionally, look at that as a lot of medical services claims (as much as 30–40%, as indicated by a KPMG study) don't consent to true requests. Programming bots can undoubtedly recognize those exemptions and consequently save numerous superfluous installments. By performing more precise information section, bots additionally further develop charging effectiveness.

5.7.3 FURTHER DEVELOP INCOME CYCLE CAPACITIES (NEW UNDERSTANDING ARRANGEMENT DEMANDS, PATIENT PRE-APPEARANCE AND APPEARANCE, GUARANTEE DISAVOWALS, CHARGING)

Pay cycles much of the time incorporate many code changes which can be problematic for the structure. Mechanical cooperation computerization (RPA) is the right measure to ensure predictable variety to these adjustments, and accordingly an overall knowledge. These legitimate cycles get a lift from data digitization and from the robotization of drawn-out tasks like records payable.

5.7.4 SUPPORT LARGE-SCALE IMPLEMENTATION OF HEALTH PLANS

By further developing patient records the board (because of innovation highlights like high precision and minimization of human mistake), RPA in medical care brings about following more normalized, patient-explicit courses toward patients' ideal wellbeing destinations.

Mechanized cycles permit all the more convenient distinguishing proof of patients who stray from the plans, and thus, they make it simpler to bring them in the groove again. Indeed, chronic a long way-off statement of individual records at the degree of populace wellbeing without mechanical cycle computerization includes exceptionally enormous expenses. As per Peter B. Nichol, "Wellbeing plans and suppliers are finding programming specialists as a practical option in contrast to upgrading or supplanting stages."

5.7.5　IMPROVEMENT OF THE CONSIDERATION CYCLE

This is another positive outcome of the recently referenced constant record checking made possible by additionally created data assessment with RPA. Exploring careful proportions of facts works on the chance of more unique stop activates all round uniquely designed treatment techniques. Besides, experts who don't have to actually follow mainly a tremendous deal of facts due to the fact the bots can do it can location extra time in managing, and giving human assistance to, their patients. This is one more portrayal of an incredibly humanist aftereffect of RPA sending—the way that advancement makes people really matter.

5.7.6　ADVANCED ADMINISTRATIVE CONSISTENCE

Mechanical cycle computerization (RPA) guarantees that all interaction steps are followed, recognizable and reported, just as deliberately coordinated in all around organized logs. This works with the medical care organization's semi-perpetual preparation to manage outer reviews, even surprising ones.

5.8　HOW IS THE PANDEMIC ACCELERATING RPA ADOPTION?

Coronavirus has impacted each industry on the planet, and as per Bain and Company, it has sped up the mechanization initiatives. In the medical services area, COVID has generally upset the day-to-day tasks. A new World Health Organization review shows that practically 50% of the nations overviewed (49%) announced disturbance in treating diabetes. Forty-two percent of the respondents noticed issues in treating malignant growth. While 31% couldn't work well in cardiovascular crises. The UK alone has a build-up of 12 million dropped arrangements because of the pandemic. There's nothing unexpected that scientific offerings suppliers are going to robotization.

5.8.1　SOME HEALTHCARE ORGANIZATIONS ARE STILL CAUTIOUS

Robotized medical services frameworks have numerous advantages—they diminish manual responsibility, work on understanding experience and increment worker fulfillment. In any case, facilities are not hurrying into computerization. Furthermore, here's the reason:

Introductory venture and support costs: Automation needs huge assets. As indicated by Deloitte, one bot can cost somewhere in the range of $5,000 and $15,000. Directors need to compute when the benefit from RPA will take care of the underlying speculation. And surprisingly after the arrangement is set up, sellers can in any case charge for upkeep of their item and programming refreshes.

Plausibility of disappointment: In keeping with considers, 30-half of mechanical robotization projects fall flat. The reason for disappointment can be specialized or human-related. For instance, directors probably won't think about their facility's IT foundation while putting resources into programming robots or may essentially robotize some unacceptable assignments. The human factor assumes an extra part: workers may not select in if directors don't consolidate computerization drives with social change.

Dread of occupation substitution: Healthcare suppliers can decline to work with RPA programming or even endeavor to undermine it out of dread of losing their positions. Despite the fact that RPA can't duplicate human psychological capacities, it can supplant numerous normal assignments.

5.8.2 Improving on Claim Processing

Guarantee preparation is an extended interaction of contributing information, directing assessment and taking care of requests. Other than losing time, managerial specialists can make administrative mistakes identified with information section or consistence. As per consideration, four out of ten submitted medical services claims don't follow official requests.

Robotization in the clinical business can fix this issue. This is what RPA bots can do:

- Spot consistence-related exemptions
- Perform a more exact information passage
- Improve correspondence with outsiders
- Reduce guarantee preparing costs
- Eliminate overabundances by running external the standard working hours

5.8.3 The State of Arizona Employs RPA Solutions

The province of Arizona is a magnificent illustration of RPA arrangements by and by. It gives healthcare coverage through Care1st Health Plan Arizona. First and foremost, the state utilized a blend of 15 Excel macros to manage guarantee preparing and patient enlistment. This strategy was moderate and inconsistent. As a method of embracing advanced mechanics robotization in medical care, the province of Arizona sent bots that generously sped up guarantee looking after. An assignment that required as long as ten hours to execute was diminished to one. Moreover, this innovation can handle 20 cases each moment, which is right around multiple times quicker than utilizing macros.

5.8.4 Extricating Patient Data while Remaining Compliant

RPA clinical apparatuses can naturally look for right persistent records in the medical clinic's data set and send them to the correct individual inside the association.

This sort of robot would require a human representative's accreditations to get to the organization. Medical care laborers can prepare RPA programming to remove report metadata, including creation date and ID quantities of electronic clinical record archives. Workers acquire this data by checking the properties of pertinent records and envelopes. In the wake of "watching" this activity a few times, bots can impersonate people. Some RPA programming suppliers call this "work area mechanization." Moreover, programming robots ensure all the cycles are detectable and archived in organized logs. This game plan makes it simple to manage outer reviews, regardless of whether they are startling.

5.9   HOW IS THE PANDEMIC ACCELERATING RPA ADOPTION?

Coronavirus has impacted each industry on the planet, and as per Bain and Company, it has sped up the robotization drives. In the medical services area, COVID has generally upset the day-to-day activities. A new World Health Organization review shows that practically 50% of the nations studied (49%) revealed disturbance in treating diabetes. Forty-two percent of the respondents noticed issues in treating malignancy. While 31% couldn't work well in cardiovascular crises. The UK alone has an accumulation of 12 million dropped arrangements because of the pandemic. There's not anything unexpected that medical care providers are going to computerization.

5.9.1   Some Healthcare Organizations are Still Cautious

Robotized medical services frameworks have numerous advantages—they diminish manual responsibility, work on tolerant experience and increment representative fulfillment. Nonetheless, facilities are not hurrying into computerization. Furthermore, here's the reason:

Introductory speculation and upkeep costs: Automation needs critical assets. As indicated by Deloitte, one bot can cost somewhere in the range of $5,000 and $15,000. Supervisors need to compute when the benefit from RPA will take care of the underlying venture. And surprisingly after the arrangement is set up, sellers can in any case charge for upkeep of their item and programming refreshes.

Probability of disappointment: According to contemplates, 30-half of mechanical mechanization projects fall flat. The reason for disappointment can be specialized or human-related. For instance, supervisors probably won't think about their facility's IT framework while putting resources into programming robots, or may basically robotize some unacceptable errands. The human factor assumes an extra part: workers may not pick in if administrators don't join mechanization drives with social change.

Dread of occupation substitution: Healthcare suppliers can decline to work with RPA programming or even endeavor to attack it out of dread of losing their positions. Despite the fact that RPA can't imitate human psychological capacities, it can supplant numerous normal errands.

5.9.2   Simplifying Claim Processing

Guarantee preparing is an extended interaction of contributing information, leading assessment and taking care of requests. Other than losing time, regulatory specialists can make administrative blunders identified with data section or consistence. As indicated by considers, four out of ten submitted medical care claims don't follow official requests.

5.9.3 Automation in the Clinical Business Can Fix This Issue. This Is What RPA Bots Can Do

- Spot consistence-related exemptions
- Perform a more exact information passage
- Improve correspondence with outsiders
- Reduce guarantee handling costs
- Eliminate overabundances by running external the standard working hours

5.9.4 The State of Arizona Employs RPA Solutions

Practically speaking, the territory of Arizona is a brilliant illustration of RPA arrangements. It gives medical coverage through Care1st Health Plan Arizona. Before all else, the state utilizes a blend of 15 Excel macros to manage guarantee handling and patient enrolment. This strategy is moderate and questionable. As a method of embracing mechanical technology robotization in medical care, the territory of Arizona conveyed bots that substantially accelerated assure taking care of. An assignment that required as long as ten hours to carry out was diminished to one. Besides, this innovation can deal with 20 cases each moment, which is right around multiple times quicker than utilizing macros.

5.9.5 Extricating Patient Data While Remaining Compliant

RPA clinical instruments can naturally look for right persistent records in the clinic's data set and send them to the correct individual inside the association. This sort of robot would require a human worker's accreditations to get to the organization. Medical care laborers can prepare RPA programming to extricate report metadata, including creation date and ID quantities of electronic clinical record archives. Representatives acquire this data by checking the properties of significant records and envelopes. Subsequent to "watching" this activity a few times, bots can mirror people. Some RPA programming suppliers call this "work area computerization." Moreover, programming robots ensure all the cycles are discernible and reported in organized logs. This plan makes it simple to manage outer reviews, regardless of whether they are surprising.

Here's the reason RPA is significant for measure mechanization:

- Improve speed, quality and efficiency RPA bots can be prepared to attempt un-shrewd and tedious undertakings quicker and more precisely than people at any point could.
- Get more worth from large information numerous associations are producing such an excess of information that they can't handle every last bit of it. There are numerous chances to acquire experiences from this information and drive more noteworthy efficiencies. RPA is obviously fit to help parse through enormous data sets, both organized and unstructured, assisting associations with figuring out the information they are gathering.

- Free up representatives for more important errands RPA offers the chance to let loose representatives to chip away at more significant undertakings. Abandoning drawn-out and tedious work, representatives can take up the positions of things to come, maybe up skilling to execute mechanization and AI to accomplish more prominent results.
- Become more versatile to change—Recovering from the disturbance brought about by COVID-19 includes associations turning out to be more dexterous and deft in managing change. Versatility and flexibility are fundamental to defeating current and future difficulties. RPA assists associations with accelerating measures while lessening costs; guaranteeing associations are prepared to manage disturbance and change.

5.10   ROBOTIC CYCLE AUTOMATION (RPA) IN HEALTHCARE

RPA in medical care alludes to the mechanical innovation that is utilized to cover routine assignments in the business. They need minimal expense and are simple to embrace— and owing to these attributes, they've become a well-known advancement in recent times. Among its commonplace applications, Deloitte eyewitnesses notice composing notes about patients, planning therapy solutions, gathering measurable information and following past clinical history.

5.10.1   Data Management

At the section level, RPA bots can play out all the fundamental data-related tasks, including separating information, ordering records and discovering vital contact subtleties. The commonsense use cases here can be different. RPA can be applied for enlistment cycles, holding and working with clinical records or making dreary information passages.

Along these lines, the innovation works with scaling, because of the normalization of its standard methodology. Likewise, it can upgrade the adequacy of each interaction coming, making opportune admonitions and alarms if something turns out badly. This expands the precision of each move made in the medical services organization to the greatest.

5.10.2   Appointment Scheduling

With regard to arrangement booking, robotic process automation (RPA) uncovers the limit of its adequacy. For instance, RPA bots can consequently filter each specialist's accessibility, working hours and abilities and offer time allotments for patients depending on their side effects. Afterward, it will be simpler for specialists to work with patients as the robot as of now gathers all the fundamental data ahead of time. The product additionally oversees updates and sends takeaway notes.

5.10.3   Managing Claims

RPA in the medical services industry aids higher cases the executives. Here, the product helps insurance agencies in distinguishing misrepresentation and mistakes. With sped up information handling, the issues of incorrectness and deferrals are tackled. RPA robots will fill the information as per industry guidelines, oversee claims and track the viability. Therefore, the two payers and safety net providers will set aside time and cash spent on fixing botches.

5.10.4 OPTIMAL CARE DELIVERY

RPA dispenses with the requirement for manual information sources. It builds the precision of determinations because of the capacity to contrast the authentic wellbeing information and the most extreme data about the condition of a specific patient. The entire cycle is completely programmed, which altogether cuts the time and exertion spent on deciding suitable treatment.

With prescient investigation procedures, you can break down cases naturally and get a bunch of arrangements, care alarms and admonitions. Thus, it's feasible to work even with unstructured information, as RPA will look for the required snippets of data by watchwords.

5.10.5 HOSPITAL MANAGEMENT

RPA adequately oversees modern Big Data in your clinic: the undertaking that requires colossal exertion and assets whenever took care of physically. The product acquaints upgraded information investigation with adapt to this difficulty all the greater viably. Thus, it works with the work with patient records and empowers information-driven dynamic. Plus, the innovation adds to the advancement of regulatory cycles, upgrades in monetary administration and synchronization of conveyance measures.

5.11 ADVANTAGES OF RPA IN HEALTHCARE

1. Accurate robotization: Mechanical Process Automation robotizes the cycles as well as builds their precision.
2. Cost decrease: The fundamental advantage of RPA is its capacity to speed up key medical services measures so that you don't have to spend numerous assets on finishing them later on.
3. Staff advancement: RPA in medical care gives great relief to your laborers. Since robots complete every one of the unremarkable cycles rather than people, your representatives will have more opportunity to accomplish key work and feel more connected with and more joyful.
4. Transformational force: Once progressed, robots can bring substantially more worth than essentially duplicating and moving records quicker. For this situation, RPA can lead to discussions, structuralize tumultuous information and settle on complex choices. Thus, your association turns out to be more inventive.

The Checklist to Getting an Effective RPA Solution:

1. Determine your necessities: Set your assumptions and characterize the ideal effect. The more clearly you see your objectives, the more exact programming you'll get.

2. Get an expert engineer: The more abilities your IT expert has, the better arrangement this individual will convey.
3. Stay required during the venture: Numerous requestors will in general put in a request, discover an agent and disregard the venture till they get the outcome. For RPA improvement, it's urgent to stay in contact with your group and take an interest in project plan and the executives to get what you truly need.
4. Consider setting up a checking framework: The RPA programming is exceptionally subject to human conduct—as, synchronize the activities of your representatives with the responses of robots. Else, you can lose information due to a sole manual secret phrase update.
5. Include RPA into your current programming framework: Each innovative arrangement should work flawlessly as a piece of the entire interaction to be genuinely viable. That is the reason to join RPA so that it enables the innovations you as of now have.

5.11.1 Further Developing the Medical Services Cycle

Suppliers gather huge measures of information from their patients every day from individual data to treatment cycle subtleties. With the assistance of RPA programming, medical care associations can extricate and enhance patient information all the more easily. In speaking with other computerized situation, RPA programming can control gathered information to produce investigation that offers clinical staff significant bits of knowledge to help them make more exact findings and offer custommade medicines to patients.

5.11.2 An Outline of RPA in Medical Services

The medical services area in any nation is a complex yet basic framework to the prosperity of the residents. It's contained medical gear, scientific coverage, scientific preliminaries, and that's only the top of the iceberg.

Overseeing data identified with clinical applications, outsider gateways, booking applications, human asset applications, undertaking asset arranging (EPR) and radiography data frameworks is regularly a tough errand. The reconciliation across these frameworks is similarly difficult. It requires patients, insurance agencies, specialists and numerous different partners to guarantee consistent consideration conveyance.

Keeping a harmony between the rising number of people looking for care and desk work engaged with clinic measures requires a more productive and precise administrative center cycle. An administrative center in a medical services organization incorporates the organization and the care staff. The administrative center staff work on administrative consistence, record support and give IT, settlement, freedom administrations.

Progressed programmed arrangements like mechanical cycle mechanization help medical services associations lower functional expenses, increment functional proficiency and diminish human mistake in information preparing. RPA frameworks can handle data in a record time and address difficulties coming about because of human mistake

Promptly after the COVID-19 pandemic being announced, mechanical interaction mechanization (RPA) turned into a fundamental everyday instrument for some medical services suppliers, as opposed to a drawn out yearning.

For associations unexpectedly troubled with staff deficiencies and different requirements brought about by the COVID-19 emergency, the advantages of computerization were presently not absolutely about making cost efficiencies, yet about building flexibility.

Assignments like charging, preparing claims, overseeing patient records and other dull, tedious undertakings are appropriate to programming bots, as they can help save time for medical services staff to go through with patients.

> Medical care associations have these exceptionally muddled kinds of charging and claims measures that include adjusting the cases and ensuring that codes are appropriately applied. That is simply such a lot of manual information passage, checking of records, now and again checking of real sweeps.
>
> (Roberto Valdez, Director of Cyber security Automation and Risk Advisory Services at Kaufman Rossin, a warning firm for US organizations)

An investigation by Gartner completed in May 2020 found that while just 5% of medical services suppliers had been putting resources into RPA before the pandemic, 50% arrangement to do as such in the following three years.

This bodes well given that RPA has arisen as a nearly customized answer for the difficulties of COVID-19. Consider staff who regularly do administrator obligations having to hole up for about 14 days, alongside the expanded quantities of patients going through an emergency clinic. Robotization can deal with the expansion in quiet booking and arrangement retractions. "These are the regions where RPA has the greatest chance and can have the greatest effect, and the pandemic will keep on driving reception," Valdez says. One model is Indian protection firm ICICI Lombard, which currently utilized cloud-based computerization driven by AI to deal with credit only cases mentioned by medical clinics. Information like specialist's conclusions and treatment plans are taken care of into the calculation, which concludes whether to acknowledge the case, while an AI calculation settles on the case sum. A cycle that once required 4 hours to be completed physically, presently requires 90 seconds.

5.11.3 SETTLING COVID-EXPLICIT DIFFICULTIES

RPA has likewise been sent to take care of explicit issues that have emerged because of the pandemic. Coronavirus testing requires a considerable measure of monotonous administrator, remembering finding and enlisting patients for the framework, effectively naming test packs for assortment and examination and logging test results, sometimes into a public information base containing a great many different passages. Also, there is an enormous interest for tests that has made excesses in numerous nations, while social removing rules stay set up for staff.

Back in March, RPA programming firm UiPath dispatched a task with the Mater Misericordiae University Hospital in Dublin, giving their Infection Prevention and Control office with free preliminary robot licenses to handle test results. UiPath says this is saving 3 hours out of each day, which medical caretakers can rather spend on the pandemic reaction. In the UK, toward the beginning of the pandemic, 750,000 individuals reacted to the Government's call for volunteers to help with COVID-19 endeavors, in the space of only two days. Prior to taking up their jobs, everyone should have been enrolled and gone through historical verifications. While HR would normally play out these capacities physically, the sheer quantities of individuals requiring preparing implied that utilizing mechanization would be a lot quicker and more effective. Programming firm Credential was delegated to robotize the sign-up and check interaction of specialists, attendants and paramedics, enrolling a large number of clinicians in only days rather than months.

5.11.4   Moderate Appropriation, Up to This Point

Coronavirus has truly featured the advantages of RPA, yet for what reason did it take a worldwide pandemic for it to turn out to be generally received? "There's in reality some valid justifications for this," Valdez says. "Some of it identifies with the affectability of patient data, some of which is driven by consistence," however at that point additionally great patient consideration and stewardship of patient records too.

> HIPAA consistence gives specific kinds of requirements, and because of that, medical services associations have been reluctant to reevaluate particular sorts of functionalities generally, however it's gotten considerably more typical. Then, at that point when they do re-appropriate, they regularly need to confine it topographically, for example restricting it to inside the United States. In specific conditions, they will seaward, however with truly severe and inflexible limitations.

Another explanation is absolutely down to clinicians' propensities. Such a great deal what is done is in an actual space with patients, where medical care professionals are composing notes by hand, or checking graphs. Indeed, even today there are still huge loads of paper engaged with this framework. There are medical services record frameworks that are progressively ready to give great applications and interfaces to advanced forms of these, yet whatever has been prepared in the cake and utilized for an extensive stretch of time is still very pervasive.

The other part is similarly just about as straightforward as execution and framework. At the point when you go through the work, cost and season of incorporating a framework like mechanical cycle robotization, then, at that point it's presumably for your potential benefit to simply progress forward that pathway of computerized change. That incorporates the utilization of the cloud and Software as Service sort of frameworks. Mechanization across the whole worth chain is something that is proceeding to increment.

RPA is the quickest developing portion of the product market, and Valdez predicts that suppliers of other, customary kinds of programming will consider this to be a chance. "You have undertaking RPA stages like Blue Prism and UiPath that are simply massively fruitful, however at that point these conventional, enormous endeavor organizations and programming engineers are seeing that there's some portion of the overall industry accessible," he says.

Suppliers of electronic wellness statistics (EHR), client relationship managers (CRM) and endeavor asset arranging programming (ERP) are incorporating renditions of RPA into their current stages. Indeed, even organizations like Workday, which delivers an immense ERP framework, and Microsoft, are incorporating RPA into their foundation. We will see an increment in that. It'll make a decent aggressive cross-over which benefits medical services and which eventually benefits the patient too.

5.12   CONCLUSION

This pandemic is changing the lives of whole world including young talents. As this COVID 19 spreads rapidly across the world, it creates fear and mental stress. Many people are losing their lives due to incompatibility of beds and oxygen. The easiest ways to triumph over a majority of these influences is to design robots for our future use. The technologies in this modern world are trying to decrease the death rates all over the world. The RPA learns and mimics human duties, and it is going to be easier and very rapidly developing generation. The presentation of RPA in COVID fields will be extremely helpful and the finest try to diminish the risks of specialists and their lives. RPA is a brilliant mechanization association that offers an assortment of freedoms to work on excellent, increment manipulate and adaptability and opens a wide extent of computerization prospects. Notwithstanding, an unmistakable RPA vision and methodology, looking after the execution method and the working version is essential for development.

BIBLIOGRAPHY

Davenport, T., Guha, A., Grewa, D., & Bressgott, T. (2019, October 10). How Artificial Intelligence Will Change the Future of Marketing.

DeBrusk, C. (2017, October 24). Five Robotic Process Automation Risks to Avoid. *MIT Sloan Management Review.*

Evolution of Robotic Process Automation (RPA): The Path to Cognitive RPA. (2018, August 29). AIMDek Technologies.

Fersht, P. (2012, October 25). Robotic Automation Emerges as a Threat to Traditional Low Cost Outsourcing. *HfS Research.*

Frey, C.B., & Osborne, M. (2013, September 1). The Future of Employment: How Susceptible Are Jobs To Computerisation? *Oxford Martin School.*

High, P. (2019, October 21). Gartner Announces Top 10 Strategic Technology Trends for 2020. *Forbes.*

Hodson, H. (2015, March 31). AI interns: Software Already Taking Jobs from Humans. *New Scientist.*

Hugos, M.H., & Hulitzky, D.. (2010). *Business in the Cloud: What Every Business Needs to Know About Cloud Computing.* John Wiley & Sons, Inc. ISBN: 978-0-470-61623-9

Karamouzis, F., & Da Rold, C. (2014, January 24). Predicts 2014: Business and IT Services Are Facing the End of Outsourcing as We Know It. *Gartner Research.*

Karthiga, M., Nandhini, S. S., Tharsanee, R. M., Nivaashini, M., & Soundariya, R. S. (2021b). Blockchain for Automotive Security and Privacy with Related Use Cases. In *Transforming Cybersecurity Solutions Using Blockchain* (pp. 185–214). Springer, Singapore.

Karthiga, M., Sankarananth, S., Sountharrajan, S., Kumar, B. S., & Nandhini, S. S. (2021c). Challenges and Opportunities of Big Data Integration in Patient-Centric Healthcare Analytics Using Mobile Networks. In *Demystifying Big Data, Machine Learning, and Deep Learning for Healthcare Analytics* (pp. 85–108). Academic Press.

Karthiga, M., Sountharrajan, S., Nandhini, S. S., & Kumar, B. S. (2020, May). Machine Learning Based Diagnosis of Alzheimer's Disease. In *International Conference on Image Processing and Capsule Networks* (pp. 607–619). Springer, Cham.

Karthiga, M., Sountharrajan, S., Nandhini, S. S., Suganya, E., & Sankarananth, S. (2021a, March). A Deep Learning Approach to Classify the Honeybee Species and Health Identification. In *2021 Seventh International Conference on Bio Signals, Images, and Instrumentation (ICBSII)* (pp. 1–7). IEEE.

Kochan, T. (2019, January 10). It Is Not Technology That Will Steal Your Job. *The Irish Times*.

Lacity, M., Willcocks, L.P., & Craig, A. (2015). Robotic Process Automation: Mature Capabilities in the Energy Sector. *LSE Research Online Documents on Economics*.

Lacity, M. C., & Willcocks, L. (2015, June 19). What Knowledge Workers Stand to Gain from Automation. *Harvard Business Review*.

Maddox, T. (2019 October 21). Top 10 Technology Trends for 2020 Include Hyperautomation, Human Augmentation and Distributed Cloud. *TechRepublic*.

Menear, H. (2020, May 17). Gartner Tech Trends 2020: Developing the Multiexperience. *Technology*.

Ojala, A., & Helander, N. (2014). Value Creation and Evolution of a Value Network: A Longitudinal Case Study on a Platform-as-a-Service Provider. *47th Hawaii International Conference on System Sciences* (pp. 975–984). doi:10.1109/HICSS.2014.128.

Overby, S. (2012, November 16). IT Robots May Mean the End of Offshore Outsourcing. *CIO*.

Panetta, K. (2019, October 21). Gartner Top 10 Strategic Technology Trends for 2020. *Gartner*.

Ratia, M., & Myllärniemi, J. (2018, July 6–7). Beyond IC 4.0: The Future Potential of BI in the Private Healthcare. *Proceedings IFKAD 2018*. Delft, Netherlands (pp. 1652–1667).

Saari, E., Käpykangas, S., & Hasu, M. (2016, October 7). The Cinderella Story—A Skilled Worker's New Chance in the Digitalization of Services. *Finnish Institute of Occupational Health*.

Sountharrajan, S., Nivashini, M., Shandilya, S. K., Suganya, E., Banu, A. B., & Karthiga, M. (2020b). Dynamic Recognition of Phishing URLS Using Deep Learning Techniques. In *Advances in Cyber Security Analytics and Decision Systems* (pp. 27–56). Springer, Cham.

Sountharrajan, S., Suganya, E., Karthiga, M., Nandhini, S. S., Vishnupriya, B., & Sathiskumar, B. (2020a). On-the-Go Network Establishment of IoT Devices to Meet the Need of Processing Big Data Using Machine Learning Algorithms. In *Business Intelligence for Enterprise Internet of Things* (pp. 151–168). Springer, Cham.

Suganya, E., Sountharrajan, S., Shandilya, S. K., & Karthiga, M. (2019). IoT in Agriculture Investigation on Plant Diseases and Nutrient Level Using Image Analysis Techniques. In *Internet of Things in Biomedical Engineering* (pp. 117–130). Academic Press.

Testing: Getting Started with Process Automation (RPA) for SAP. (2019, September 29). Miracle Software Systems.

Willcocks, L. P., Lacity, M., & Craig, A. (2015, October 30). The IT Function and Robotic Process Automation. *LSE Research Online Documents on Economics*.

6 Newfangled Immaculate IoT-Based Smart Farming and Irrigation System

Ms. P. Divya

Bannari Amman Institute of Technology, Erode, India

D. Palanivel Rajan and K. S. Kannan

CMR Engineering College, Hyderabad, India

D. Yuvaraj

Bannari Amman Institute of Technology, Erode, India

Balachandra Pattanaik

Wollega University, Nekemte, Ethiopia

CONTENTS

DOI: 10.1201/9781003181668-6

6.1   INTRODUCTION

Agriculture plays an energetic role in our country. An economy that will enable the nation to increase GDP growth is an essential factor in agriculture. Most of the geographical area in our nation has occupied agricultural land, and even this sector produces more jobs.

6.1.1 THE NEED FOR AGRICULTURAL COMPUTATIONAL INTELLIGENCE

Most industries, including agriculture, use intelligence technologies. The deployment of intelligence in agriculture saves excess water, decreases human resources, lowers the cost of goods, etc. It helps to grow healthier crops and monitor soil conditions, which provides the conditions with data for farmers to decide. Agricultural intelligence integrates into android mobiles, IT platforms, and IoT technology, which provides the knowledge to help the farmers find a more effective way to improve the products and goods from the different changes that happen because of climatology.

6.1.1.1 Use of Robotic Applications in Agriculture

The manual tasks make productivity plodding, repetitive tasks for farmers. To overcome this problem, these kinds of tasks are automated by agricultural robots, enabling them to concentrate on improving the overall output yields. The most popular robots which are used in the field of agriculture are:

- Crop Seeding
- Crop Monitoring and Analysis
- Fertilizing and Irrigation
- Harvesting and Picking
- Crop Weeding and Spraying
- Thinning and Pruning
- Autonomous Tractors

These are some of the applications where we use robotics in the field of agriculture. Using this farmer can able to improve the size of yields and minimize the wastage crops. One of the most common robotic applications in agriculture is harvesting and picking—for example, a robotic designed to pick sweet pepper that can encounter several hindrances [1]. In harsh conditions, a vision system must need to assess for locality and mellowness of the pepper, which has dust, changing light intensity, temperature fluctuations, and wind-generated movement.

But we need some sophisticated vision systems for choosing a pepper. To delicately grip and position the pepper, the robotic arm must traverse in environments with just as many hindrances. This technique is significantly dissimilar to picking and positioning the metal component in an assembly line. The robotic arm must be flexible and accurate enough in a complex environment that can't damage the peppers when they are picking.

The robot used for harvesting and picking is becoming more popular among farmers. Still, there are hundreds of innovative ways for the agricultural industry to use robotic automation to increase production yields. The demand for food is now overtaking the farmland available, and it is up to farmers to close this distance.

6.1.2 CHALLENGES FOR IMPLEMENTING IoT IN AGRICULTURE

In applying IoT technology, the agriculture sector faces three significant barriers that are mentioned in subsequent sections [2].

6.1.2.1 Weak Farm Internet Connectivity

Most farms are placed in remote areas where internet access is not sufficiently effective to allow the data to broadcast at high speed. Moreover, crops, canopies, and other physical barriers might obstruct contact lines. These factors increase the cost of data broadcast and slow the adoption of precision technologies in agriculture.

6.1.2.2 Elevated Prices for Hardware

To gather data on farm conditions, farmers currently depend on sparsely distributed sensors in the network. In addition to the physical constraints of these sensors, farmers need to continue to rely on fewer advancements in farming technologies for limiting their productivity.

Instead, tethered eye helium balloons are used in areas where drone use restrictions occur, including government regulation, low life of a battery, and high costs. To overcome this situation, FarmBeats used uncrewed aerial vehicles (UAVs) to increase the spatial coverage and establish precise measurements. This aerial sensor generates a stream of continuous images for farm conditions used to progress the data obtained by the sensors from the ground.

6.1.2.3 Disrupted Cloud Access

Like every other IoT framework, Farm Beats built a cloud computing system using Microsoft's Azure Platform. Sometimes internet access will not be good enough to stream big data sets for analysis to the cloud. This approach also harms the cloud link.

6.1.3 Smart Agriculture Practices

There is a diverse range of technologies used in intelligent farming to replicate the scope of the activities carried out by farmers, growers, and other stakeholders in the field [3]. The following seven application areas are designed for intelligent farming:

- Fleet management-farm vehicle monitoring
- Arable cultivation, farming of large and small fields
- Monitoring of Livestock
- Greenhouses and stables-indoor farming
- Monitoring for storage: water tanks and fuel tanks

6.1.4 Benefits of Adopting New Technology in Agriculture

Agriculture plays a vital role in generating a profit. This sector has faced many challenges and improvements in various agricultural methods and techniques in the last few years. Nowadays, inorganic fertilizers, the consumption of reduced pesticide amounts, several tractors and machinery are used. In addition to increasing agricultural productivity and reducing costs, this input availability shows the necessitated usage of natural resources and processes. There are a lot of advantages available when agriculture is implemented with the use of modern technology.

6.1.4.1   Adoption of Technology in Agriculture

In Agriculture, modern technologies are used in different cultivation fields, like herbicides, pesticides, fertilizers, and seed improvement. In recent years technology has proven to be highly needed in the agricultural sector. In the current situation, without the ideas, farmers are facing problems about the crops where it grows, to overcome this issue, the possible way is only through biotechnology in agriculture. Genetic engineering has made it possible to enhance certain trains for other genes in plants or animals. Such things will increase crop resistance to pests and droughts. Farmers are in a role, via technology, to captivate the process of improving the efficiency and quality of production.

There is a restriction on how the process can speed up with new technology adoption in agriculture, which will be accelerated because accelerating the idea requires a great deal of awareness and understanding of some of the elements that affect farmers' decisions when implementing modern technology in farming. The size of the land, the expense, and the technological advancements play a significant role in economic factors that influence agricultural technology adoption. The level of education, age, communal alliance, and gender of farmers are some social factors that can influence the farmer's likelihood of adopting modern agricultural technologies.

The implementation of new technology factors in agriculture is a concern. Small-scale farmers can face problems during the implementation in internal and external challenges. Section 6.1.4.2 illustrates the importance of modern technology in agriculture.

6.1.4.2   Technology Use of Agriculture

There are various types of technological improvements in agriculture, which include the following ones.

6.1.4.2.1   Farm Robots

One of the problems faced by farmers nowadays is the need for labor with satisfaction. There is a rising in the cost of labor. This issue can be solved by simplifying and introducing hybrid harvesters and planters. Some significant elements in agriculture during harvesting time include sowing the seeds, planting, watering the plants, cutting the weeds, applying fertilizers and pesticides, inspecting the soil and collecting the yield, cutting the weeds, and killing the plague and insects that damage the crops [4]. It ensures the usage of modern agricultural technology, which can produce large quantities of food in the shortest possible time.

Developing sprayers and autopilot tractors does not require any driver, that GPS has been used. Such technology is significant, which promotes better and more efficient farming techniques.

6.1.4.2.2   Sensors for Crops

In agriculture, the utilization of pesticides and fertilizers remains a significant challenge, especially when determining which nutrient works best for various plants. Using a crop sensor, farmers can make it easier to apply fertilizers and pesticides effectively, as much as crops need. Crop sensors are designed to dictate the application machinery about the amount and time of the resource needed for a given crop. In such cases, variable rate technology is used.

The "green seeker sensor" was developed at Oklahoma State University. This intelligent machine reads the needs of a plant and then applies precisely about the appropriate amount of herbicide fertilizer. It is a device that uses sensors to let the plant tell us what it needs. It will shine a light on plants at red and near-infrared wavelengths, absorbing that light and reflecting some light into sensors. The sensor tests the amount of light reflected from the plant and calculates the quantity of fertilizer required by the plant. By using this, we can understand the amount of fertilizer needed by the plant and the system that takes signals from the fertilizer amount to the plant [4].

6.1.4.2.3 *Application of GPS in Field Documentation*

In agriculture, implementing GPS is the most common technology. Modern agriculture includes the use of GPS to document the condition of the farmland. The yields from a given farm are easy to determine the log and record the application rates. These things are beneficial; farmers may rely on the collected and registered data for reference when making some decisions.

6.1.4.2.4 *Biotechnology*

Genetic engineering and the method of changing genes of a given crop are often referred to as biotechnology. Genetic modification is most commonly carried out to improve the confrontation for certain crops to agricultural inputs, like herbicides. Reduced farm inputs ensure that the farmer saves the expense in agricultural resources as well.

New agricultural technology aims to accomplish two important goals: profit in the economy and improved production. However, to achieve these goals, farmers need to have the idea of modern farming and technology. Thus, it is important to be vigilant about the objectives and set for different agricultural technologies to be implemented. Farmers should look at how to spread fertilizer, irrigation, theatre, intensive tillage, monoculture, and application of other instruments.

6.1.5 State of IoT in Agriculture

By automating processes and offering valuable data that could affect decision-making, IoT technologies can dramatically transform agricultural enterprises.

6.1.5.1 Data Collection and Processing

Farming, like any other enterprise, requires rich data. A profitable corporation that does not care about the predictions of figures and analysts is scarcely possible to imagine. The combination of IoT and software systems allows data to be collected, processed, analyzed, and visualized in a way that provides easy access to each stakeholder and enables them to track the current business situation.

6.1.5.2 Control over Risks

Because IoT devices collect loads of data while algorithms operate to transform it into something useful, it is possible to monitor business risks. Let's assume that the algorithm has estimated an average number of crops that will harvest this year, month, day, or whatever. Then, farmers should modernize their distribution scheme to prevent financial losses caused by unsold production.

6.1.5.3   Business Automation

With the assistance of collected and analyzed data, business processes can be easily automated. For instance, farmers want to know the average number of crops, when each form is harvested, etc.

6.1.5.4   Quality Improvements

Sensors for IoT agriculture may detect and report on various types of problems in the field of research (soil issues, humidity levels, and even livestock diseases). Timely avoidance of problems will produce the results in the quality of a better end product.

6.2   IMPACT OF INTELLIGENCE

In India, the agricultural sector is detrimental, and crop production is decreasing day by day. To overcome this issue, we need to identify and implement a new way to increase production in the agricultural industry. The agriculture field can gain importance by using intelligent technologies such as IoT, computational intelligence, big data, cloud-based services, and GPS [5]. To achieve higher precision in agriculture, we need to analyze crops, which is necessary to transform live field data and automation techniques in farming to achieve high productivity in fields. It can be achieved by introducing intelligent techniques in farming. This section discusses the scope, role, and advantages of computer intelligence and IoT.

6.2.1   Role of Intelligence in Agriculture in the Modern Era

Today, most agricultural start-ups are adapting to the AI-enabled approach to increase their productivity in agriculture. The Market Study Report states that at the end of 2025, the global market size of Artificial Intelligence (AI), which involves in the field of agriculture, is projected to reach US$ 1550 million. AI-empowered techniques could be applied to identify diseases or climate change earlier, making the farmers react smartly. AI offers more productive ways necessary for making the crops to be grown, harvest, and sell.

AI implementation focuses on testing faulty yields and enhancing the potential for safe crop production. Most agro-based companies produce more effectively in the production with the implantation of artificial intelligence technology. This technology is also used for weather forecasting and disease or pest detection in applications such as automatic system adjustments. Artificial intelligence will thus enhance crop management practices by introducing new technologies that will help in agriculture. AI strategies have the potential to address the problem which is faced by the farmers, such as a change in the climate, insect and weed infestation, which makes decreases in the yields.

6.2.1.1   Effects of Artificial Intelligence in Agriculture

AI technology is becoming an inseparable tool for every agronomist and farmer in today's agricultural industry. The AI technology quickly corrects the issues while advising practical action needed to resolve the issue. In short, Precision Agriculture uses technological advancements such as robotics, GPS guidance, control systems, drones, autonomous vehicles, and sensors to make farming more accurate and controllable [22].

6.2.1.1.1   Weather Data Forecast

The study of information gathering helps farmers to take a precaution by comprehending and learning with AI. AI helps farmers stay up to date with weather forecasting in an advanced way. The predicted information allows the farmers to raise the yields and income without risking the crop. An intelligent decision can be taken on time by enforcing such procedures.

6.2.1.1.2   Crop and Soil Quality Monitoring

The use of AI is an important way of performing or tracking potential soil defects and nutrient deficiencies. AI detects potential faults with the image recognition method by images collected by the camera. It analyzes patterns using agriculture flora which is developed with the aid of AI and deep learning system. By understanding the soil defects, plant pests, and diseases, these AI-enabled applications are helpful.

6.2.1.1.3   Decrease Pesticide Usage

To manage weeds, farmers can implement the AI techniques like computer vision, robotics, and machine learning. With AI techniques, farmers can keep the analyzed information and spray the chemicals where it's needed. By this method, farmers can decrease the number of spraying chemicals in the entire field. It will reduce the utilization of herbicides in the field compared with the number of chemicals usually sprayed.

6.2.1.1.4   AI Agriculture Bots

The farmers can find effective ways to protect their crops from weeds by implementing AI-enabled agricultural bots. It will help the labor to overcome the difficulty in fields. Using this technique, farmers can harvest more crops in a higher volume and perform more speed work than human workers in the agricultural sector. It helps to track the weed and spray it by using computer vision. It is one of the ways for farmers to identify an efficient way for protecting their crops [6].

6.2.2   Scope of Internet of Things in Agriculture

An IoT-related artificial intelligence (AI) uses drone over agricultural lands. Drones help to check the farm lands' current status and send the data via image Processing. The drones are used in soil field research, crop tracking, and health assessment of field crops. The AI technique assists in the progress of the crops. It helps by all odds the farmlands. Overall, IoT can control the entire farmlands. Therefore, the plants growing in an unhealthy way can be observed, and then an automated treatment will be given to those plants if required.

The high cost and less alertness of smart devices are restricting their use in intelligent farming. The often changes in cost also affect the usage of smart devices, which increases the overall price of the end product, where it prohibits the use of new technology by farmers. Farmers can quickly move to Internet of Things (IoT), which increases the efficiency of food products in the global market and reduces human workload, time, and expense. Traditional methods are carried out and tested every day in the manual calculation of climate variables. This technology helps the farmers analyze climate, humidity, temperature level, and soil fertility by keeping the past information. By implementing a crop monitoring system that supports the farmers to connect with the farm at any time in a remote location. IoT incorporates several devices without human involvement to transfer the data collected into information. In this field, challenges will start in cultivating crops on a farm and making the best possible price for the product's sale to the customer.

In recent decades, climate shifts and erratic rainfall due to global warming have influenced the agriculture cycle. Following traditional techniques, the production of raw materials is still steadily declining. Because of these impacts, many intelligent technologies have been adopted by farmers to make agriculture bright. Computational intelligence for remote monitoring is built into technologies such as IoT to help vast numbers of applications produce their performance more efficiently and accurately.

It offers the combination of different sensors and substances that, without human intervention, can make communication directly within the devices. The physical devices, like sensor devices, observe and collect the details. The IoT's zenith increases steady global linking with people, artifacts, sensors, and services. Farmers have had their conventional techniques for cultivating crops according to their needs and economic structure since time immemorial.

6.2.3　IMPORTANCE OF INTELLIGENCE IN AGRICULTURE PROCESS

The information such as temperature, precipitation, wind speed, solar radiation, and assessments of historical data details are provided for each part of the agricultural ecosystem. Soil sampling, agricultural robotics, crop monitoring, and predictive analytics are the three key categories in agriculture used for data collection by farmers and stored in the farm management systems, which supports better processing and analysis [7]. By knowing the sense of strengths and weaknesses of their soil, farmers can avoid the cultivation of defective crops and optimize safety potential in crop production. With this, the growth of AI technology increases the efficiency of agro-based companies. Although it will not replace human farmers' jobs, it will provide an efficient way of growing, harvesting, and selling crops.

By identifying a swarm in an annoying parcel of a field, farmers can implement this technique. Using the plant's image with an algorithm, the farmer can make a good control for saving maize from grasshoppers, and a robotic lens zooms into the yellow flower of a tomato seedling. By implementing this, farmers can get more from the soil while enabling resources to be used more sustainably.

A See & Spray robot allegedly leverages a computer vision to identify and spray plants weeds accurately. The camera and sensors images are collected, and the robots can analyze various weeds to spray correct herbicides needed for the invasion field.

6.2.4 Benefits of Computational Intelligence

AI has a significant effect on intelligent water management (drip or sprinkler irrigation), variety selection (heat/drought/stress-tolerant), conservation tillage (minimized loss of moisture and soil), accurate nutrient supplement (save input cost), and farm management software are five main measures to boost agricultural productivity for achieving a paradigm shift in a production aspect. From drones, the AI-enabled cameras capture near-real-time images of the entire farm and analyze those images to identify issues and come up with possible improvements. Robots, for instance, would help farmers tackle the problem of a declining workforce and allow them to work more effectively while saving labor money.

Advanced robotic systems and on-farm data collection can also prepare and harvest plants, increasing crop yields. In agriculture, AI bots (Agri-robots) can act like today's combined harvesters that make harvest crops at a higher volume rate than human workers [8]. Machine visions are used to their full benefit to assist in the control, weeding, and spraying used for in-vivo agriculture. Reduced costs include the benefits of robotics adoption in Indian agriculture: agri-bots being used in many regions in India tending to crops, harvesting, weeding, etc., can reduce the cost of fertilizers and remove human labor and attract young people. The future of agriculture, especially Indian agriculture, will be witnessed by many innovations through AI adapted to regional requirements and diversity.

6.3 OVERVIEW ON SMART FARMING

Smart farming is a new way that highlights the use of new technologies that are expected to harness its growth and incorporate more robots and AI techniques into agriculture. It includes the Big Data phenomenon, large quantities of data with a wide variety that can be collected, processed, and used to make a decision [23].

6.3.1 Various Steps in the Farming Process

Farm automation entails intelligent farming techniques that make farms more productive and automate production processes with crops or livestock. Many firms are working on this robotic engineering with drones, autonomous tractors, robotic harvesters, automatic irrigation, and seeding robots. While these innovations are relatively recent, many traditional farming companies have introduced farm automation in their industrial processes.

In general, before embarking on any crop production, farmers anywhere need to consider the aspects such as which crop to sow in the field, tillage process, fertilization program, demand and price for the product, etc. Such factors are relatively prevalent. For instance, if any one needs to cultivate wheat, there is no need to distillate on that aspect, but other aspects will remain. Figure 6.1 shows the steps carried out during farming.

6.3.1.1 Crop Selection

During the crop selection, farmers should consider the total revenue of the crop planted by them, such as grains, millet, fruits, and flowers. They need to remember

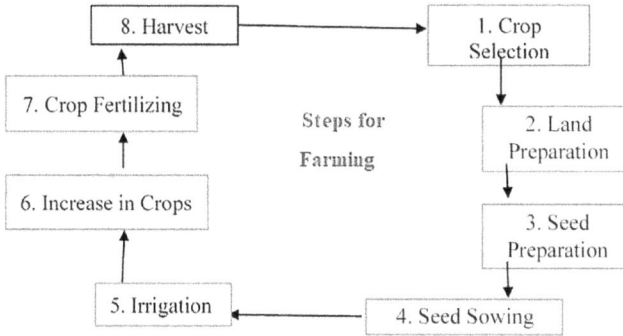

FIGURE 6.1   Various steps followed in farming.

some aspects like which crop will give the best market price when harvested at that time of sowing? What is the future demand for that crop in the market? Is the crop appropriate for the land and the climate? What fertilizers, quantities of such fertilizers, and timings of application would be required? Is the requisite irrigation infrastructure available or has to be built, and at what cost? The above sounds were overwhelming. Still, such information is readily available for some short time to acquire knowledge in firsthand or through make a discussion with the government agriculture department and through an ever-present internet.

6.3.1.2   Preparation of Land

The next step is needed to choose the crop for the cultivation appropriate for the land he had. The philosophy is always that we should grow good land at the least amount and time for cultivation. Farmers are currently using tractors for tilling operations almost everywhere in the world. In some places, due to local conditions, manual and animal tillage are still in practice. There are some methods available for land preparation.

6.3.1.2.1   Land Preparation Method—Tillage Practices

The first move or physical effort allows the soil to cut, plowing breaks, and overturns the soil. The total phosphates and organic manure used at this stage of the process are mixed with the soil. The next stage is the harrowing of tillage for smoothing and pulverizing the soil to a shallow depth. Weeds are drawn out of the soil here, and fertilizers are combined with the soil as well. This move would also increase the humidity of the soil. Leveling is the third stage in raising land value and making provisions for water application, uniform soil moisture.

After these processes, farmers need to decide which sort of tillage is needed for the crop. This tillage must be done with various kinds, such as deep tillage, shallow tillage, or tillage still on earth with earlier crop residue that gives a rollover to make a softer seeding. The soil is less dense, and increased rooting would need to be done. Then uprooting all old crop weeds and residues, irrigation water can go deeper quickly for good growth. It allows sun rays and air to land, which improves the soil and kills some viruses.

Tillage, if conducted manually, becomes expensive and more. But some alternative methods are available to complete conservative tillage, which depends on farmer, crop, cost of energy, and land. Minimum tillage and nil tillage are the alternative methods available in farming.

6.3.1.2.2   Seed Preparation
The farmers need to select the high-quality seeds, which help for the fast germination. Moisture is a must in soil tilled by any technique; compacting the soil around this seed also guarantees needed moisture. For specific crops, seeding must be done uniformly with appropriate density. Successful cultivation is not produced by overseeding. The smallest seeds are required for zero and conservative tillage. The weed control should be done to prevent the weeds from getting into the fertilizer and water intended crops.

6.3.1.2.3   Sowing Seed
Farmers need to take sufficient time for seed sowing and consider optimal conditions at the time of sowing. The most acceptable form for seed sowing is sowing seeds in the depth of soil.

6.3.1.2.4   Irrigation
The essential part of farming for a crop is irrigation. The sum of water that should be supplied to the plants is the essential task for making the crops grow in a good way to produce more yields, so the frequency of water supply must be monitored carefully during irrigation.

6.3.1.2.5   Increase in Crops
In a given area, more than the optimum number of seed sprouts are often planted. For healthy plant growth, farmers must reduce density and make the average crop growth rate under normal conditions. Farmers compare the input of crop growth rate, leaf size, and crop color with expected growth in given conditions that intervene to sustain expected crop growth. We need to consider the properties such as fertilization frequency and quantity for proper plowing time and weeding time.

6.3.1.2.6   Crop Fertilizing
Monitoring the moisture content in the soil is necessary after the seeds have sprouted and also needs to avoid any stress in water for the plants. If water has been stressed, crops become incredibly unproductive. Needed fertilizers must be applied to the crop time-to-time. A fair rule is that urea should be applied in three equal parts, at the beginning of seeding and twice after the crop has grown. Phosphates are provided once at the beginning and need to be added in later stages with potash.

6.3.1.2.7   Harvest
The last process in farming is harvesting, where the farmers need to decide the proper harvest time and process. During this process, he needs to keep a proper stock of crops and consider the transportation expense in his profit. After this, he needs to compare market rates to get a good profit.

6.3.2 Using Modern Sensors in a Farming Environment

Farmers use intelligent sensors used in precision farming to communicate to optimize crops and evolving ecological factors. Farmers need to understand their crops by positioning a sensor on a micro-scale, controlling energy, and decreasing the effect on the ecosystem. They need to map their fields accurately in this stage and keep track of areas in the proper way where it needs fertilizer and weed treatments exactly.

Now the farmers are using agricultural robots extensively due to labor shortage and need to increase a feed due to the rise of the global population.

Farmers need to keep track of their field pest population available in remote places and take appropriate action by using online cloud services and dashboards to protect their crops using new intelligent sensing techniques. To attract and catch a particular insect trap uses a pheromone lure. Then that information is sent to the farmer's smartphone or laptop. Then they can also view their field as a satellite image, counting how many pests in each trap are caught, along with data about historical patterns and the use of pesticides. To perform these kinds of operations some sensors are used on farms. In Table 6.1, we list some types of sensors and their uses.

A loss of time and productivity could occur if a farm vehicle collapses. Now, farmers can collect and monitor data from their field equipment remotely. To overcome the issues in mechanical machines, such as use of tractors and farmers are assisted by telematics equipment to alert mechanics when failure can happen by using this telematics system. The majority of agricultural equipment firms are designed.

6.3.2.1 Benefits of Agriculture Sensors

The advantages or benefits of agricultural sensors are as follows:

- They are designed to meet growing food demand by optimizing yields with nominal resources such as water, fertilizers, and seeds will be achieved by conserving energy and mapping fields.
- They are easy to use and easy to mount and also cheaper.
- They can also be used for emissions and global warming in addition to agriculture.

6.3.2.2 Drawbacks of Sensors for Agriculture

The following are the drawbacks of using Agriculture Sensors:

- Continuous internet connectivity is needed for intelligent farming and IoT technology.
- In the developed countries such as INDIA and other parts of the world, this issue is the most significant challenge.
- Basic infrastructure specifications such as intelligent grids, traffic networks, and mobile towers are not available everywhere. It further hampers the growth.

TABLE 6.1

List of Sensor Types with Description

S. No	Type of Sensor	Description
1	Location Sensors	We can use location sensors to measure the latitude, longitude, and altitude of any position within the area. The GPS satellite is used for this purpose.
2	Optical Sensors	The properties of soil use light to measure near-infrared, mid-infrared, and polarized light spectrums for various light reflectance frequencies. These sensors are mounted in vehicles or drones to collect soil reflectance and crop color data for processing. By using this sensor, moisture content, soil clay, and organic matter can be measured.
3	Electro-Chemical Sensors	This sensor can mount especially on designed sleds, which support collecting the process and map with soil chemical data. It provides the necessary data for precision farming to identify the nutrient level of soil and pH. By using an ion-selective electrode, determination of pH measurements are carried out. The specific ions like nitrate, potassium, or hydrogen are sensed by this electrode and provide the details.
4	Mechanical Sensors	Farmers have used this sensor for measuring the compaction of soil-applied to variable degrees of compaction. In this, the device cuts soil and records force calculated by load cells or pressure gauges. When the sensor cuts ground, resilience forces resulting from soil cutting, fracturing, and displacement are tracked.
5	Dielectric Soil Moisture Sensors	In this moisture level, dielectric constants are monitored by these sensors, an electrical property that depends on the moisture content of the soil. In farm humidity sensor is used in conjunction with rain gauge stations to observe the conditions of soil moisture when vegetation levels are limited.
6	Air Flow Sensors	Airflow sensors measure the permeability of soil air. When on the move at particular locations or dynamically, measurements can be made. The desired output of pressure will push a certain level of air to the ground at a certain depth.
7	Electronic sensors	This sensor is mounted in tractors and various field vehicles to check the operations of the equipment. Cellular and satellite data networks were used to relay the information automatically to an e-mail of individuals. Then field supervisor will be able to obtain that data for analysis.
8	Self-contained sensors	Agricultural weather stations are monitored by this sensor which locates in different areas around rising fields. There is a mixture of sensors that is suitable for local crops and climate at these stations. The calculation is performed at scheduled intervals with data such as air temperature, soil temperature at various depths, rainfall, leaf wetness, chlorophyll, the velocity of wind, dew point temperature, wind speed, relative humidity, sunlight, and atmospheric pressure.

TABLE 6.2
Advantages of Automation Over Traditional Farming

S. No.	Traditional Farming	Smart Farming
1	Here farmers need to keep laborers for handling crops	Here farmers can monitor their farms with the help of advanced technologies
2	Need more time to perform farming processes	Need less time and simplified in farming processes
3	Need to invest for more cost for laborers and equipments	More cost-efficient farming
4	Quality will be less	Better quality

6.3.3 ROLE OF AUTOMATION IN AGRICULTURE AND ITS ADVANTAGES OVER TRADITIONAL METHOD

The implementation of automation in agriculture will have more benefits than a traditional approach [9]. In Table 6.2, we discuss some advantages of using automation over traditional farming (Table 6.2).

6.3.4 APPLICATIONS OF IoT IN AGRICULTURE

Some of the applications of IoT in agriculture are listed in the following sections [10].

6.3.4.1 Climate Conditions

In the context of climate change, the decision of the farmers to settle on cultivation in agriculture plays a significant role. For example, if incorrect climate knowledge is provided to farmers, which substantially deteriorates the quantity and efficiency of crop production, then to overcome this, IoT provides solutions to check weather conditions in real time within and outside the fields where sensors are located.

6.3.4.2 Precision Farming

It is the most popular far-reaching IoT application that allows more precise and controlled farming practice by implementing a livestock observation [11]. It processes information that sensors generate and responds to it accordingly. Farmers generate the data and analyze it to make wise and fast decisions with the help of sensors in precision farming.

6.3.4.3 Smart Greenhouse

To create intelligent greenhouses, IoT will follow weather stations to perform automatically and based on instructions it will adjust climate conditions. Using IoT in the greenhouse will minimize human intervention and make the whole process cost-effective and precise. To create a modern and cheap greenhouse by using a solar-powered IoT sensor.

6.3.4.4 Data Analytics

The traditional database system does not support the data obtained from sensors; instead, cloud-based data storage and an end-to-end IoT platform are used in Smart Agriculture Framework to overcome this problem. Here sensors are a primary source for gathering information in IoT environment and these use analytic software for processing the data and translating into useful information like weather condition, livestock condition, and crop condition. Predictive analytics provides an idea for making better harvesting-related decisions.

6.3.4.5 Agricultural Drones

Agricultural operations have almost been revolutionized by technological developments, and using agricultural drones is a trend of disruption. The crop health evaluation, planting, crop spraying, and field analysis are done using land and aerial drones. Drones with thermal or multispectral sensors will make changes used in irrigation processes where it's needed. Sensors indicate health and measure vegetation index once the crops start to develop. Smart drones have eventually minimized the environmental impact [14].

6.4 PROCESS INVOLVED IN IRRIGATION

6.4.1 INTRODUCTION ABOUT IRRIGATION AND ITS IMPORTANCE

Irrigation is the process of supplying water artificially for crop cultivation. Water supply limits make a continued population growth step up the search for water conservation measures in urban agriculture. The efficiency of water use depends on the performance of irrigation technology and management practices [21]. Water supply limits and continued population growth have stepped up the search for water conservation measures in urban agriculture. The efficiency of water depends on the performance of irrigation technology and management practices.

The different forms of irrigation methods include wells, reservoirs, lakes, canals, and dams. It provides the moisture level needed for the growth of crops, germination, and other related functions. The following points bring out the importance of irrigation:

- Insufficient, uncertain, and irregular rainfall causes agricultural uncertainty. The rainy season is limited to only four months, and the remaining eight months are dry. Rainfall is scarce and unreliable in many regions, even during the monsoon.
- Productivity in irrigation land is significantly higher than the productivity in unirrigated land.
- Successful implementation of high yielding will greatly enhance agricultural production.

6.4.1.1 Irrigation Types

Various types of irrigation are practiced for making improvements in the yield of a crop. Irrigation systems may be built for dissimilar types of soils, ecosystems, plants, and resources. Table 6.3 lists some types of irrigation methods followed by farmers.

TABLE 6.3
Different Types of Irrigation Process

S. No	Type of Irrigation	Description
1	Irrigation of the surface	Water is divided by gravity across the land. No pump will be used.
2	Localization of irrigation	A network of low-pressure pipes does the supply of water for each plant.
3	Irrigation of sprinklers	Distribution is done from a central location by high-pressure overhead sprinklers or by moving platform sprinklers.
4	Irrigation drip	Drops of water are produced close to the roots of plants. It's rarely used because it needs more maintenance.
5	Centre for pivotal irrigation	By using a sprinkler system, water moves in a circular pattern.

6.4.1.2   Irrigation Methods

Irrigation is done by two different methods: (i) Traditional Irrigation and (ii) Modern Irrigation.

6.4.1.3   Traditional Irrigation Method

In this, irrigation is done manually. Farmers pull water from wells or canals on their own or use cattle to pour water into farmland. This method can vary from region to region. The main benefit of using this method is cheap. However, due to the uneven distribution of water, its efficiency is poor, and there are very high chances of water loss.

6.4.1.4   Modern Irrigation Method

The modern method overcomes the disadvantage of the traditional method and contributes to the proper use of water. The modern method consists of two systems:

6.4.1.5   System of Sprinklers

The pump is connected to the pressure-generating pipe, and the water is sprinkled through the pipe nozzles. Here it sprinkles the water over the crop and evenly helps in the distribution of water. This method is highly recommended in regions where water scarcity is present.

6.4.1.6   Drip System

Drip irrigation relies on exceptionally safe water. The creek pump easily clogged in holes with the tiniest grain of mud sucked up, which meant that soil around the clogged holes had no water. Then we need to add a severe filter and adjust it periodically if you intend to install drip irrigation for city water. Drip irrigation needs many hoses and, if you practice crop rotation, they have to be transferred every year. Drip irrigation is intended for crops in rows, not for beds. Drip irrigation is costly [12].

6.4.1.7 The Need for Automation in Irrigation

In the conventional system, the farmer has a different irrigation schedule for different crops. In the Precision Agriculture, a key component is the intelligent irrigation process. It helps to avoid wastage of water and improve the growth of the better-quality crops in fields by (a) providing water at the right time, (b) minimizing runoff and waste, and (c) specifying a level of moisture in soil precisely to identify irrigation requirements at any location. Substituting automated valves and systems for manual irrigation also removes human error factors (e.g., failing by turning off the valve after irrigation in the field). It is instrumental in power saving, time, and valuable resources.

6.4.2 ROLE OF ROBOTICS IN THE IRRIGATION PROCESS

New technologies are playing an increasingly important role in improving the productivity of the farming industry. Both irrigation machines and technology have essential roles. Intelligent systems and robots are now used in addition to irrigation systems, reducing workforce defects and saving energy and time. In general, existing irrigation systems are divided into pressure and gravitational systems; sprinkler and drip irrigation systems are included in the pressure group, and furrow irrigation is typically included in the gravitational system. Water plays an essential role in agriculture, which is proven by the fact that most of Iran lies in a desert climate. An automated device also meets the need without the permanent presence of humans during monitoring in the growing season. But human errors have resulted in operators' errors or delays in taking the necessary steps, reducing productivity. Thus, with the development of waterworks and different instruments, contortion would be the under-pressure units allocated to most sources. Automatic irrigation scheduling is valuable for optimizing the efficiency of water usage concerning manual irrigation for direct calculation of soil water. The estimate of plant evapotranspiration is an alternative limitation for estimating crop irrigation needs. ET is influenced by weather parameters, including relative humidity, solar radiation, temperature, speed of the wind, and crop variables, like growing phase of crop, variety and plant density, soil properties, pests, and disease control [13].

6.4.3 IRRIGATION DEPLOYMENT IN SMART FARMING

In smart farming, irrigation is one of the most complicated ways in which we need to use water helpfully without any loss. With the aid of the sensor and intelligence robot, we discuss another method of irrigation process in this section. Figure 6.2 explains the elements which are found in this technique.

6.4.3.1 Details for Land

A sensor is filled with the collect data on soil moisture, temperature, and humidity, send the details to the Raspberry-pi kit, and store them in the central database.

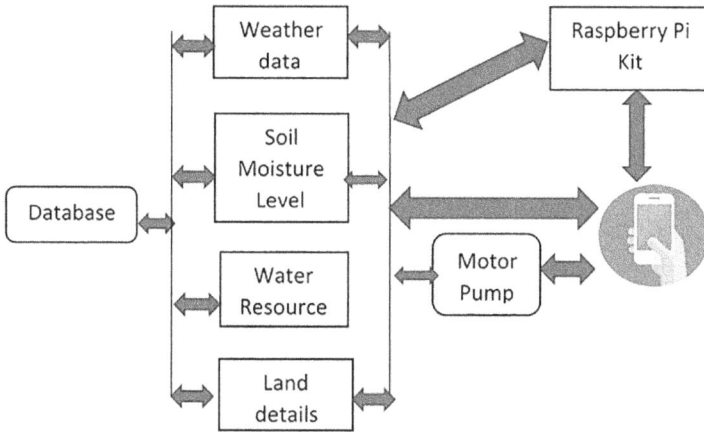

FIGURE 6.2   Simple irrigation process architecture diagram.

6.4.3.2   Data on the Weather

The weather data must be obtained from the different internet sources, which contain temperature, humidity, and rainfall online. This information is stored in the central database that is useful for finding soil moisture and is used in the algorithm for prediction.

6.4.3.3   Moisture Level of Soil

The soil moisture level is based on the details of crops grown in the field and the weather details from the weather data. The random forest and agglomerative clustering algorithms are used to find the soil, and the water level needed for crops is also predicted.

6.4.3.4   Recourse to Water

The water level is stored in the database and is linked to the set of motor pumps used to make water flow in the field. This motor pump is linked to the Wi-Fi module sensor kit for a start and stop, and it also has water availability and control information. The water will also be supplied to the field from a remote area using a mobile device after water-level availability has been collected.

Figure 6.2 shows a simple, intelligent framing irrigation method using IoT devices and the intelligence algorithm. Water is used well by using an IoT in an irrigation system and promoting water loss in fields. Farmers manually and automatically carry out this irrigation with the help of a mobile device whenever they are in a remote area.

6.5 SECURITY ASPECTS IN SMART AGRICULTURE

The agriculture sector plays a vital role in human society. Despite the benefits of this development, many security threats can seriously affect the agricultural domain. According to the latest research, the world will need to produce a portion of food for more than 70% in 2050, relative to current production, and also need to feed constantly increasing population in the Planet, which is expected to hit almost 10 billion by 2050 [16]. For meeting these needs, innovation needs to be used in the agriculture sector to increase crop cultivation. While the new standard is anticipated to be Agriculture 4.0, physical threats and risks are also critical factors in this field, preventing widespread acceptance and adoption. Some challenges, such as environmental patterns, have historically remained the same over the years, while others are synonymous with the overwhelming evolution of technical solutions. Calicioglu et al. [18] offers a short and detailed list of physical risks to agriculture, which stated some issues such as (i) changes in climate, (ii) quick growth in global population, urbanization, and aging, (iii) progressive systems in the production of food which make an impact on farmland and farmer, and (iv) diseases and pest security for using an IoT in agriculture.

Intelligent communication technology and IoT integration bring new threats and hazards to the global market regarding ICT security. In a complex and distributed cyber-physical environment, possible attacks on the different intelligent agricultural systems may lead to serious security problems. These types of attacks and threats lead to integrated severe enterprise disruptions. Precision Agriculture and smart farming use innovative technology and remote administration for their stakeholder to increase the mechanized agricultural landscape, where a new threat in this specific field is closely linked to similar threats that occur in other industries. Such as cybersecurity, data integrity, and data loss are often related to these risks [19,20] and many emerging vulnerabilities, which may also potentially lead to catastrophic consequences because the Precision Agriculture sector uses heavy machinery linked online [20]. Some of the areas where risks are possible are:

- Risks in IoT vulnerabilities in agriculture and IoT layer threats
- Vulnerabilities and hazards in modern agriculture of equipment cybercrime and agriculture cyber security.

6.5.1 SECURITY FOR USING AI IN AGRICULTURE

The innovation of intelligent devices with sensing and acting capabilities makes the IoT framework readily available. As the number of devices connected to the network is massive, much data is created. It is a daunting job in an IoT environment to process and perform computing. So, along with some other new technologies, artificial intelligence comes as a rescue to fix the IoT security issue. As shown in Figure 6.3, IoT and AI will combine to enhance the system's analysis, improve operational performance, and improve accuracy. The AI could allow IoT to calculate enormous quantities, unstructured data, heterogeneous data in real time, making the device practical [17]. IoT and AI integration with some simple features is shown in Figure 6.4.

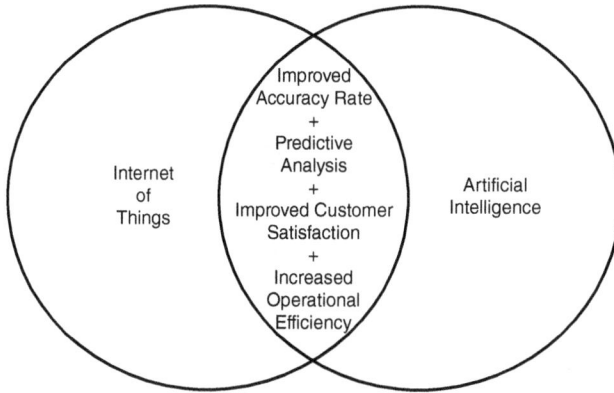

FIGURE 6.3   The popular IoT and artificial intelligence features.

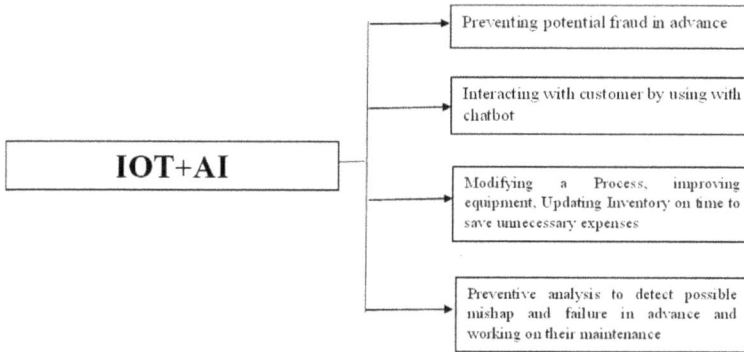

FIGURE 6.4   IoT and AI integration and their essential characteristics.

With the rapid growth in global populations, there is a need to increase farm yield to become essential. Furthermore, the problem has been compounded by the limited utilization of expected sources, such as water and field, which make a yield of some staple crops gradually. Changing composition of the agricultural workforce is another impending problem for the agricultural industry. In addition, agricultural labor has also decreased in most regions. The adoption of internet connectivity in agricultural practice has resulted from the declining agricultural labor force to minimize manual work in the field.

6.6   FUTURE TRENDS IN SMART AGRICULTURE

Smart farming implementation is based on IoT, allowing growers and farmers to reduce waste and increase production, ranging from the amount of fertilizer used by farm vehicles. It allows resources such as electricity and water to be used well. IoT provides a solution for smart farming by using a device like a sensor (light, humidity, temperature, soil moisture, crop health) to track crop fields and automate irrigation systems. The farmers from anywhere can track the condition of their field. Based on this data, they can also choose manual or automated options for taking appropriate actions [15]. For example, if soil moisture level decreases, the farmer will deploy sensors to start irrigation. When comparing with the traditional method, smart farming is highly effective. The IoT can change agriculture procedures in specific ways, and farmers need to recognize that IoT will be a driving force by cost-effectively increasing agricultural production. Since the market is still developing, companies eager to join still have enough opportunity to produce agriculture.

6.7　CONCLUSION

In agriculture, the new age uses Industry 4.0 concepts and assets that promote a transition in Agriculture 4.0, which involves smart farming. The increasing sophistication of methods used in producing and manufacturing agri-products contributes to even more significant potential threats in different ways. Multiple "backdoors" have opened with stakeholders by making the agriculture ecosystem bright and linked, and protection specialists have also advised working together to chunk them entirely. To make safety and efficiency, all new possible solutions are offered to farmers. In conclusion, a new approach needs to (i) reduce cost, (ii) save time, (iii) increase confidence, and (iv) less risk. To characterize it as effective, they are persuaded by the newly proposed system to be secure, safe, and functional, improving efficiency and providing added value, making agricultural stakeholders implement a new way of working. Considering the above, it becomes apparent that consolidation of emerging technologies in the critical agricultural field is an immense task that needs to be tackled step by step and only through the successful participation of directly affected stakeholders in security-preserving activities and investments in the supply chain.

REFERENCES

[1] https://bigdata.cgiar.org/three-major-obstacles-for-iots-in-agriculture.
[2] https://www.agrifarming.in/smart-farming-in-india-challanges-techniques-benefits #Application_Areas_of_Smart_Farmi://ng.
[3] https://www.farmmanagement.pro/modern-agricultural-technology-adoption-and-its-importance/.
[4] Mohiuddin, S.M., "Agricultural Robotics and Its Scope in India", *International Journal of Engineering Research & Technology*, 4(7), July 2015, 1215–1218.
[5] https://customerthink.com/the-role-of-artificial-intelligence-in-agriculture-sector/.
[6] https://mahtabrasheed.wordpress.com/2012/11/14/steps-a-farmer-performs-and-what-information-is-required-at-each-step/.
[7] https://myknowledgebase.in/7-steps-for-successful-agricultural-practice-by-farmers.
[8] https://www.agritechtomorrow.com/article/2019/02/smart-sensors-in-farming/11247.
[9] https://steemit.com/agriculture/@owoblow-steemit/importance-of-automation-in-agriculture-automation-of-agriculture-is-the-future.
[10] https://www.biz4intellia.com/blog/5-applications-of-iot-in-agriculture/.
[11] https://www.economicsdiscussion.net/essays/irrigation-importance-sources-development-and-limitation/2108.
[12] Sature, M. J., et al. "Irrigation Sprinkler Robot." *International Journal of Advance Research in Science and Engineering*, 7(3), April 2018, 377–382.

[13] Venkanna, et al. "Mood Automated Irrigation System Using Robotics and Sensors." *International Journal of Scientific Engineering and Research*, 3(8), August 2015, 9–13.

[14] https://www.freepatentsonline.com/and2017/0020087.html.

[15] https://www.iotsworldcongress.com/iot-transforming-the-future-of-farming/#:~:text=B Ipercent20Intelligencepercent20surveypercent20expectspercent20that,percent245perce nt20billionpercent20backpercent20inpercent202016.

[16] Demestichas, K., et al. "Security Threats Survey on Agricultural IoT and Smart Threats." *Sensors*, 20, 6458, 2020, http://doi.org/10.3390/s20226458.

[17] Mohanta, B. K., Jena, D., Utkalika, S., Patnaik, S. "The Survey IoT Security: Machine Learning Problems & Solutions, Artificial Intelligence Technology of Blockchain." *Internet of Things*, 11, 100227, 2020.

[18] Calicioglu, O., Flammini, A., Bracco, S., Bellú, L., Sims, R. The Possible Food Challenges Farming: Integrated Study of Patterns and Solutions. 2019, 11, 222 for Sustainability. [The CrossRef].

[19] Window, M. "Security in Precision Agriculture: Agricultural Device Vulnerabilities and Threats." Master's Thesis, Computer Science, Electrical and Space Engineering Department, Technology University of Luleå, Luleå, Sweden, 2019.

[20] Champion, S., Linsky, S., Mutschler, P., Ulicny, B., Reuters, T., Barrett, L., Bethel, G., Matson, M., Ramsdell, K., et al. "Threats to Precision Agriculture." 2018 US Department of Homeland Security and Office of Intelligence and Analysis Study on the Public-Private Analytic Exchange Program: Washington, DC, USA, 2020.

[21] Gravalos, I., et al. "A Robotic Irrigation System for Urban Gardening and Agriculture." *Journal of Agricultural Engineering*, 50(4), 198–207, 2019, http://doi.org/10.4081/jae.2019.966.

[22] Aydan, A. "Impact of Artificial Intelligence on Agricultural, Health Care and Logistics Industries." *Annals of Spiru Haret University. Economic Series* 19(2), 167–175, 2019.

[23] Wolfert, S. et al., "Big Data in Smart Farming—A Review," *Agricultural Systems* 153, 69–80, 2017, http://doi.org/10.1016/j.agsy.2017.01.023.

7 A Review on Haze Removal Methods in Image and Video for Object Detection and Tracking

Monika Sharma and Dileep Kumar Yadav
Galgotias University, Greater Noida, India

S. B. Goyal
City University, Selangor, Malaysia

CONTENTS

DOI: 10.1201/9781003181668-7

7.1   INTRODUCTION

Image processing is the field which is used to process the data and extract the information from the processed images. We can extract the information in the form of text, patterns or other forms that are easily understood by the machines. This is also used to identify the patterns from the images. Firstly, the images are scanned and then preprocessed. The preprocessing of images includes noise removal, skew correction and binarization. Sometimes, when photographer takes the photo from the cameras, the images get deteriorated, and sometimes their quality is degraded due to environment and weather conditions. These weather conditions may be haze, fog, mist, dust, etc. These conditions decrease the clarity of the images.

7.1.1   BACKGROUND SUBTRACTION METHOD

The basic background subtraction method is very popular and efficient technique for object detection and tracking as shown in Figure 7.1. Here, the background subtraction method computes the difference between current frame and background model in order to extract information for many computer vision applications.

Haze results in decreasing the contrast in the images. Two phenomena are air light and attenuation that result in the formation of the haze in the atmosphere as shown in Figure 7.2 [1].

The presence of airlight in the atmosphere results in whiteness, and attenuation decreases the intensity of contrast of the image [1]. Haze removal methods are used to enhance the quality of the images and restore the visibility.

While taking photograph, clarity and visibility automatically decrease if distance is increased between the camera and the desired object. Various applications such as surveillance, object detection, object tracking, etc. are based on image processing and computer vision which is a highly demanded area of research nowadays.

The suspended particles present in the atmosphere result in scattering of light, and it destroys the quality of the image. Haze results in mixing of reflected light from the image with the additional light present in atmosphere. Haze removal methods are

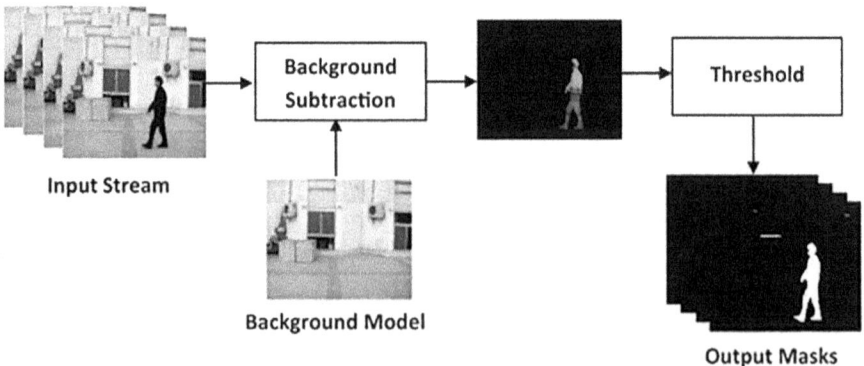

FIGURE 7.1   Moving object detection using background method.

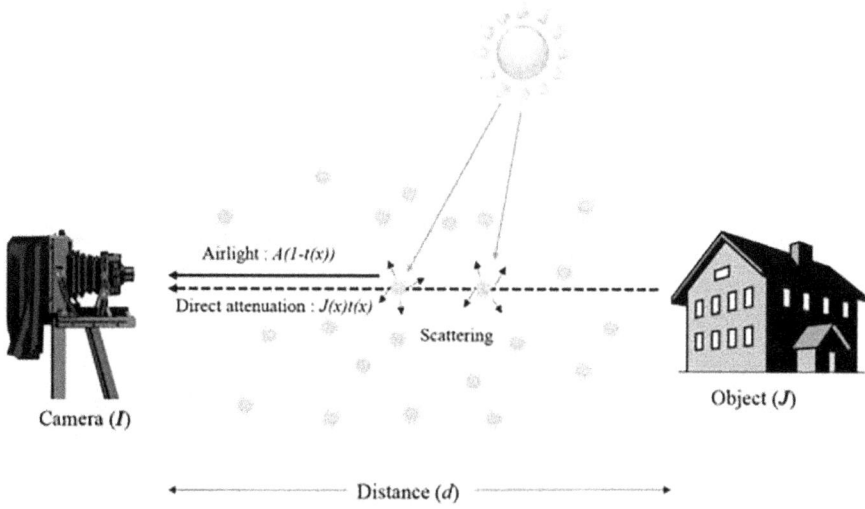

FIGURE 7.2   Haze formation [1].

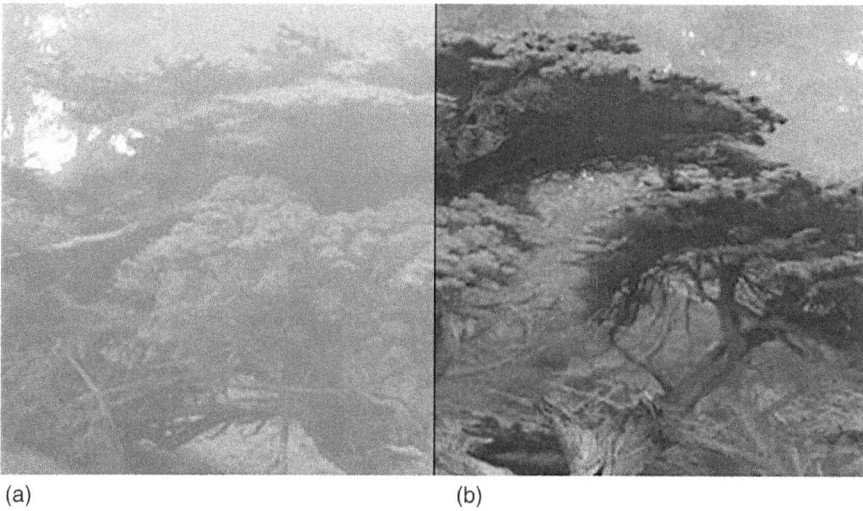

(a)   (b)

FIGURE 7.3   (a) Hazed image; (b) Dehazed image [1].

applied to any image and video to improve the reflected light from the image, from the atmosphere light as shown in Figure 7.3 [1].

There are many types of haze removal methods for the single and multiple images as shown in Figure 7.4.

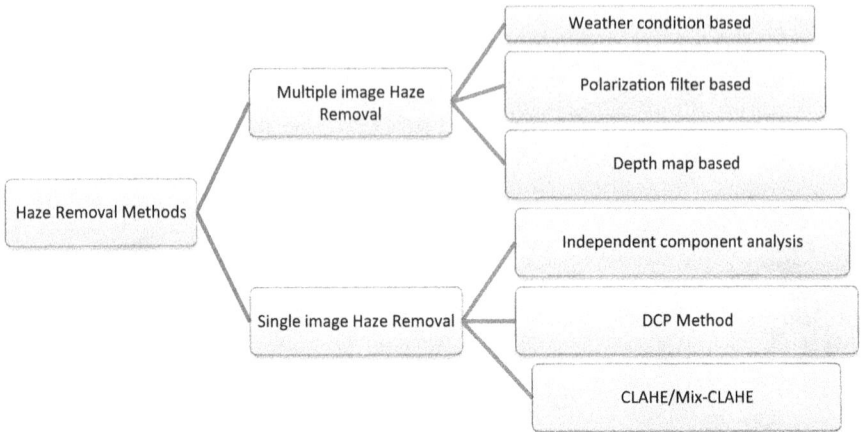

FIGURE 7.4   Haze removal methods.

7.1.2 WEATHER CONDITION-BASED METHOD

This method uses the adaptive method for dynamic scene. It uses characteristics of the medium to enhance the clarity of the scene. It takes time in giving better results.

7.2 POLARIZATION FILTER-BASED METHOD

In this method, polarization filter is attached with camera, and it is used to reduce the reflection of light present in atmosphere by rotating the polarization filter. This method is not suitable for dense Fog.

7.2.1 DEPTH-BASED METHOD

This method uses 3D projection of the scene and image. It calculates the scene depth and helps reducing contrast from the image to enhance the visibility of the scene.

7.2.2 INDEPENDENT COMPONENT ANALYSIS

This method divides the image into components and removes haze independently from each component. This method is not suitable for dense fog and dense haze.

7.2.3 DCP (DARK CHANNEL PRIOR) METHOD

DCP method calculates the atmospheric light to reduce the haze from the image. In this method, preprocessing and postprocessing steps of image processing are used.

$$\text{Dark Channel is represented by } J^{\text{dark}}\left(x\right) = \min\left(\min^{c}\left(y\right)\right) / y \in \pi\left(x\right)$$

$J^{c}(y)$ is RGB format. DCP method calculates the atmospheric light to remove haze [2].

7.2.4   CLAHE/Mɪx-CLAHE

CLAHA (contrast limited adaptive histogram equalization) method enhances the contrast to enhance the visibility of the image. This method does not affect the hue and saturation of the image and converts the RGB pixels to HSV pixels. Mix-CLAHA also uses the same methodology and is used for the enhancement of the underwater images [3].

7.3   LITERATURE

Chris Stauffer & W. Grimson (1999) [4]: This model helps in detecting people in environment and remotely control vehicles and thus provides output for real-time performance where human intervention is not involved. This method includes modeling of each pixel as a combination of Gaussians and online approximation. This method provides approximately 11–13 frames per second.

Mahfuzul Haque, Manzur Murshid & Manoranjan Paul (2008) [5]: This study gives object detection technique in which statistical framework is used with BBS (basic background subtraction) technique. This method includes PETS and WALLFLOWER datasets for surveillance taken from different angles of camera.

Dileep Kumar Yadav and Karan Singh (2016) [6]: This study uses Kullback-Leibler Divergence based on threshold and background subtraction method for object detection and tracking. CDNET dataset is used in this method.

Cao J. G. (2018) [7]: This study gives the experimental output to enhance the visuality of the image using the fractional calculus method. This method is used to find the image texture and also helps in detecting blurred images. Fractional calculus is also used to remove noise from the image. Fraction Derivative in Fourier transform is

$$D^{V}f(t) \longleftrightarrow (D_{f})^{V}(w) = (i_{w})^{V} f_{(w)} = d^{V}(w). f(w).$$

This method applies this transform in preprocessing of the image to enhance the image.

Robby T. Tan (2008) [8]: In this study adaptive method came into existence which uses single input images. The method analyzes that there is difference between the image contrast that is taken in clear weather and bad weather conditions. Also the air light factor depends on the distance between object and camera. A framework is developed using this method. Markov random field with cost function gives higher saturation values and better results. This method works less on halo effects in image.

Schechner et al. (2001) [9]: In this study polarization filter method is used to remove the haze from the static images. Polarization filter is attached to the camera. Images are captured by rotating the polarization filter of the camera. Polarized filter blocks the random light waves from passing through, creating a clearer image. Polarization filter reduces the reflections, reduces the atmospheric haze and increases color saturation in images. This method is not suitable for heavy haze and dense fog.

Zhang et al. (2012) [10]: This study develops a new method of visibility enhancement using filtering process in an image using median filter approach. This research also includes DCP method for dehazing the hazed image. Images captured in heavy fog with this method do not work so well.

Dr. Vinod Shokeen et al. (2016) [11]: This study includes DCP method, adaptive histogram equalization method (CLAHE) and mix CLAHE method for dehazing the hazed image and underwater images using DCP method.

$$\text{Dark Channel is represented by,} \, J^{dark}(x) = \min\left(\min^c(y)\right)/ y\varepsilon\pi(x)$$

$J^c(y)$ is RGB format. DCP method calculates the atmospheric light to remove haze [2].

Y. Wang et al. (2010) [12,13]: This study gives the improved DCP method where window size of atmospheric light is increased. This gives improvised version of DCP method.

R. Fattal (2008) [14]: This study proposes a new method which calculates the transmission in hazy scenes based on single input image. This approach depends on the assumption that transmission and surface shading are locally correlated. The use of these statistics allow us to manage complicated scenes with different surface criteria. This method helps to regain the contrast of image which provides the haze-free image.

K. He et al. (2009) [2]: This research model includes DCP method where, value of atmospheric light is used to remove haze from the single digital image. DCP method is completely based on statistics of outdoor images.

Li-Wei Kang et al. (2012) [15]: This study focuses on two novel image priors for haze removal for image and video. The first is a pixel-based dark channel prior and second is a bright channel prior. Here, atmospheric light is calculated by haze density analysis. This method decreases the depth of transmission map.

M. Prarthana et al. (2018) [16]: This study includes dehazing method which is used to calculate the airlight which is responsible for the formation of haze. This method repairs the transmission map dynamically and uses the bilateral filters to enhance the quality of the image.

K. He et al. (2010) [17]: This study uses novel type guided filter method to accelerate the bilateral filter to smoothen the image. The method removes the noise from the image and enhances the quality of the image.

S. Yang et al. (2013) [18]: This method uses DCP method where histogram of the image is changed by changing the contrast of the image which results in the thickness of haze in the image. So, haze is not removed properly.

Xu, Haoran (2012) [19]: This study uses the DCP method of haze removal with bilateral filtering for single images. This model estimates the refined transmission map by using DCP method for haze removal and improves the speed of DCP method.

Atul Gujral et al. (2014) [20]: This study also calculates the atmospheric light and transmission depth for removal of haze in an image. It gives better results than DCP method.

Linting Bai et al. (2015) [21]: This study is based on DCP method with downsampling and guided filter. Up sampling in this method is too simple which affects the contrast of output image and contrast of output image and also affects the dehazing of hazed image.

Zhiyuan Xu et al. (2009) [22]: This study uses CLAHE method to enhance the contrast of the image. This method converts the RGB image to HSV color image and helps in the removal of haze and fog from the image.

Hitam M. et al. (2013) [3]: This study improves the CLAHE method [22] and is called Mix-CLAHE. This new method is used to remove haze from underwater images and restore their visibility.

Falah Ibrahim et al. (2018) [23]: This study includes the various image restoration and enhancement algorithms. It also discusses various dehazing algorithms for digital images.

A. Hussain et al. (2018) [24]: This study discusses about the various types of cameras used for Traffic surveillance system. It also includes various problems occurred under extreme weather conditions such as haze, fog, mist, dust etc. during traffic monitoring and control.

7.3.1 ANALYSIS FROM LITERATURE

Reference	Author	Year	Method
[8]	Robby T. Tan	2008	• Adaptive method for single input images.
			• Less work in Halo-effects.
[9]	Schechner et al.	2001	• Polarization filter method is used for haze removal in static images.
			• Not suitable for heavy or dense fog.
[10]	Zhang et al.	2012	• DCP method used for haze removal.
			• Not worked for images captured in heavy fog.
[12]	Y. Wang et al.	2010	• Improved DCP method in which window size of atmospheric light of image is increased.
[7]	Cao J. G.	2018	• Fractional calculus method is used to enhance the visuality of the image.
			• Removes the noise from the image.
[15]	Li-Wei Kang et al.	2012	• Method uses dark channel and bright channel both priors.
			• Atmospheric light calculated by haze density.
[20]	Atul Gujral et al.	2014	• Transmission depth method.
			• Gives better results than DCP method.
[21]	Linting Bai et al.	2015	• Method based on DCP method with downsampling and guided filter.
			• Used for dehazing of image.
[22]	Zhiyuan Xu et al.	2009	• CLAHE (contrast limited adaptive histogram equalization) method is used which enhances the contrast of the image.
[3]	Hitam M. et al.	2013	• Improves the CLAHE method called Mix-CLAHE.
			• This new technique removes haze from underwater images and helps in restoration of their visibility.

7.3.2 FINDINGS FROM LITERATURE

Various techniques and methods are applied for haze removal in object detection and tracking. Maximum methods are applied and worked for the single image and multiple images for haze removal. Very less work is done toward the haze removal in videos for object detection and tracking. So, future work is to remove haze from the video frames.

7.4 CHALLENGES

Challenges in haze removal from image and video:

There are many types of challenges faced while haze removal methods are applied. While taking the image from the camera in bad weather conditions, these conditions may be mist, haze fog, dust, etc., some of the common challenges are faced due to scattered particles present in the environment. These particles diminish the clarity of the digital image. Some of the challenges are as follows:

1. Suspended and scattered particles present in the air.
2. Haze present in the environment that deteriorates the clarity of the image.
3. The presence of fog.
4. The presence of cloudy and water droplets.
5. The presence of rain.

7.5 APPLICATION AND BENEFITS

1. **Vehicle Tracking and Surveillance:** Surveillance is a very important task nowadays in day-to-day life. This helps us in security purposes. Haze removal algorithms are very beneficial in traffic surveillance system to avoid the accidents on the highways for vehicle detection and tracking in the hazy and the dusty weather, as shown in Figure 7.5.
2. **In Agriculture:** Object detection and tracking is very beneficial in agriculture also to detect the unwanted elements and objects in the agriculture field to avoid the crops damage in foggy weather. Various techniques like drones, cameras are used for tracking unwanted objects in the agricultural fields.
3. **Border Security:** In the night, vision is not clear with the dusty atmospheric conditions or we can say in the presence of haze, vision is blurred. So, for the security at borders, haze removal algorithms must work for the security persons to get the clear vision as shown in Figure 7.6.
4. **Air traffic control:** Haze removal algorithms most importantly work for air traffic channels in the dusty atmospheric conditions.

FIGURE 7.5 Vehicle tracking in haze.

FIGURE 7.6   Border security.

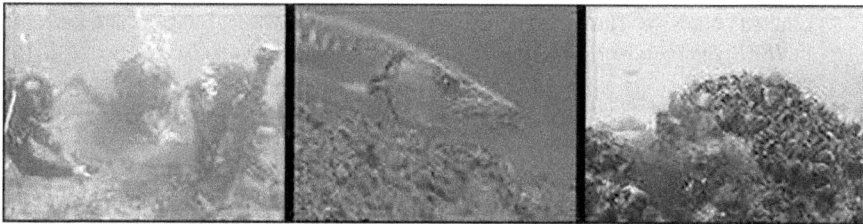

FIGURE 7.7   Underwater vision.

5. **Underwater vision and navigation:** To take clarity of pictures and images under the water and for the navigation in the presence of haze, haze removal techniques are used as shown in Figure 7.7.

7.6   RESEARCH GAP IDENTIFICATION

1. No work is done toward Noise removal in images along with the Haze and Fog removal.
2. Not much work is done on combination of CLAHE method and DCP method to improve the haze removal technique.
3. Not much work is done on removal of haze and fog in videos.

7.7   CONCLUSION

In this chapter, various haze removal methods such as DCP (Dark Channel Prior), IDCP (Improved DCP), CLAHE, and Mix-CLAHE, etc. are analyzed and reviewed. Novel defogging method gives better results for digital images. These haze removal methods and defogging techniques are used for single image and multiple images. Haze removal techniques are used to find the clarity in the images in the presence of haze particles.

7.8 FUTURE SCOPE

This chapter includes various dehazing techniques or methods applied to an image to find the clear vision in the image. Most of the research is applied to the digital images to remove the haze. Future work is to remove the haze particles from the video by extracting the frames and get clear vision for the video captured in the presence of haze in the atmosphere.

REFERENCES

[1] Aswathy, S. and Binu, V. P. "Review on Haze Removal Methods." *International Journal of Scientific and Research Publications*, 6(7), 142–145, 2016.

[2] He, K., Sun, J. and Tang, X. "Single Image Haze Removal Using Dark Channel Prior." In *IEEE International Conference on Computer Vision and Pattern Recognition*, pp. 1956–1963, 2009.

[3] Hitam, M. S., Yussof, W. N. J. H. W., Awalludin, E. A. and Bachok, Z. "Mixture Contrast Limited Adaptive Histogram Equalization for Underwater Image Enhancement." In *IEEE International Conference on Computer Applications Technology (ICCAT)*, pp. 1–5, 2013.

[4] Stauffer, C. and Grimson, W. "Adaptive Background Mixture Models for Real-Time Tracking." In *Proceedings IEEE Conference on Computer Vision and Pattern Recognition, CVPR 1999*, pp. 246–252, 1999.

[5] Haque, M., Murshed, M. and Paul, M. "On stable Dynamic Background Generation Technique Using Gaussian Mixture Models for Robust Object Detection." In *5th International Conference on Advanced Video and Signal Based Surveillance*, IEEE, pp. 41–48, 2008.

[6] Yadav, D. K. and Singh, K. "A Combined Approach of Kullback–Leibler Divergence and Background Subtraction for Moving Object Detection in Thermal Video." *Infrared Physics and Technology*, 76, 21–31, 2016.

[7] Cao, J. G. "An Image Enhancement Method Based on Fractional Calculus and Retinex." *Journal of Computer and Communication*, 6, 55–65, 2018.

[8] Tan, R. T., "Visibility in Bad Weather from a Single Image." In *Proceedings of IEEE Conference on Computer Vision and Pattern Recognition, CVPR*, pp. 1–8 2008.

[9] Schechner, Y. Y., Narasimhan, S. G. and Nayar, S. K. "Instant Dehazing of Images Using Polarization." In *The Proceedings of the IEEE Computer Society Conference on Computer Vision and Pattern Recognition (CVPR)*, vol. 1, pp. 1–325, 2001.

[10] Zhang, Y. Q., Ding, Y., Xiao, J. S., Liu, J. and Guo, Z. "Visibility Enhancement Using an Image Filtering Approach." *EURASIP Journal on Advances in Signal Processing*, 2012(1), 1–6, 2012.

[11] Shokeen, V., Bhardwaj, S. and Mishra, N. "A Study on Haze Removal Technique for Image Processing." *International Journal of Advanced Research in Computer and Communication Engineering*, 5(5), pp. 658–662, May 2016.

[12] Wang, Y. and Wu, B. "Improved Single Image Dehazing Using Dark Channel Prior." In *Proceeding IEEE Conference Intelligent Computing and Intelligent Systems (ICIS)*, vol. 2, pp. 789–792, 2010.

[13] Xiong, Y. and Yan, H. "Improved Single Image Dehazing Using Dark Channel Prior." *Journal of Computational Information Systems*, 9, 5743–5750, 2013.

[14] Fattal, R. "Single Imaze Dehazing." *ACM Transactions on Graphics*, 27(3), 1–9, 2008.

[15] Yeh, C. H., Kang, L. W. and Lin, C. Y., "Image Haze Removal via Haze Density Analysis Based on Pixel-Based Dark Channel Prior." *Proceedings of Computer Vision & Graphic Image Processing*, 2012.

[16] Prarthana, M. and Hemapriya, A. M. "Robust and Efficient Haze Removal in Image and Speed Control of Vehicle Based on Depth of Haze." *International Research Journal of Engineering & Technology (IRJET)*, 5(5), 2018.

[17] He, K. Sun, J. and Tang, X., "Guided Image Filtering." In *Proceeding International Conference European Conference on Computer Vision*, pp. 1–14, 2010.

[18] Yang, S., Zhu, Q., Wang, J., Wu, D. and Xle, Y. "An Improved Single Image Haze Removal Algorithm Based on Dark Channel Prior and Histogram Specification." In *Proceedings of 3rd International on Multimedia Technology*, Atlantis Press, pp. 279–292, 2013.

[19] Xu, H. "Fast Image Dehazing Using Improved Dark Channel Prior." In *IEEE International Conference on Information Science and Technology (ICIST)*, pp. 663–667, 2012.

[20] Gujral, A., Gupta, S. and Bhushan, B. "A Novel Defogging Technique for Dehazing Images." *International Journal of Hybrid Information Technology*, 7(4), (2014), 235–248.

[21] Bai, L., Wu, Y., Xie, J. and Wen, P.," Real Time Haze Removal on Multi-Core DSP." *Procedia Engineering*, 99, 244–252, 2015.

[22] Zhiyuan, X. "Fog Removal from Video Sequences Using Contrast Limited Adaptive Histogram Equalization." In *International Conference on Computational Intelligence and Software Engineering*, 2009.

[23] Ibrahim, F. and Rahim, M. S. M. "Current Issues on Single Image Dehazing Method." *IJCERT*, 5(2), 2018, E-ISSN: 2349-7084.

[24] Hussain, A., Maity, T. and Yadav, R. K. "Vehicle Detection in Intelligent Transport System under a Hazy Environment: A Survey." *IET Image Processing*, 14(1), 1–10, 2018.

8 Cyber ML-Based Cyberattack Prediction Framework in Healthcare Cyber-Physical Systems

P. Sathish Kumar and M. Karthiga
Bannari Amman Institute of Technology, Erode, India

E. Balamurugan
University of Africa, Toru-orua, Nigeria

CONTENTS

8.1 INTRODUCTION

Machine Learning technique helps to collect the information from the data for providing decisions in a fast and accurate manner. The data that are collected could be utilized for different purposes like descriptive, predictive, and also diagnostic purposes. Machine Learning, as known by all, is an artificial intelligence (AI) subset that helps to predict the results autonomously without any explicit coding. Once modeled, Machine Learning is capable of making predictions for any number of new data. Since the advancement in technology, data produced by the devices around us is more and the same causes problems such as difficulty in storing and in analyzing it. Machine Learning techniques could be utilized in an efficient manner to gain insights

DOI: 10.1201/9781003181668-8

from such huge big data. Different categories where Machine Learning techniques could be implemented are listed below:

- Supervised learning: Here labeled data are utilized to make decisions. An extension of such technique is ensembled learning where various small models are joined together to complete a task.
- Unsupervised learning: Here unlabeled data are utilized to make decisions with the help of patterns and various properties of the data.
- Semi-supervised learning: This technique is a combination of supervised and unsupervised techniques by gaining the advantage of both techniques.
- Reinforcement learning: This technique uses trial and error strategy to make decisions. The ultimate goal of this technique is optimization by gaining enough knowledge from the users.

Machine Learning holds a lot of applications in various areas like forecasting weather, ehealthcare, image processing, cybersecurity, etc. Cybersecurity reputation is increasing nowadays dramatically with the advancement of online applications, Internet of Things (IoT) devices and increased usage of the web. Cybersecurity is a technique that is utilized to safeguard the connected devices in the internet from attacks. Cyberspace is an area involving a lot of connected systems and devices that undergo communication. Cybersecurity is a technology introduced to safeguard this cyberspace. Cyberspace provides an interactive environment for the users to share their thoughts and communicate with other devices in the virtual space. Cyberspace has now become an imperative medium, and it is the future of the world. The security breaches concerned with the cybersecurity need to be addressed by utilizing human intelligence as well as Machine Learning techniques.

Usually, cyber-physical systems known as CPS were launched in 2006. Cyber-physical systems are intelligent, real-time, adaptive or predictable feedback systems that can be networked or distributed. CPS needs improved design tools to support design methods. These methods must support scalability, complexity control, verification and validation. CPS has applications in various fields such as healthcare, robotics, manufacturing and transportation. Figure 8.1 shows the various components of CPS. The main features of CPS are as follows:

- Integration: CPS is the integration of physical and cybernetic design.
- Limited resources: The software embeds all physical components and bandwidth and other resources, and the calculation speed is limited.
- Controlled feedback: The system maintains a high level of automation and human–computer interaction. Networked and distributed CPS includes wired or wireless networks, Bluetooth, GSM, etc.
- Complexity: CPS is strictly limited to the granularity of time and space.
- Dynamic reconstruction: CPS system has inherent adaptability.
- Reliability: Since CPS is a large and complex system, reliability and safety are very important.

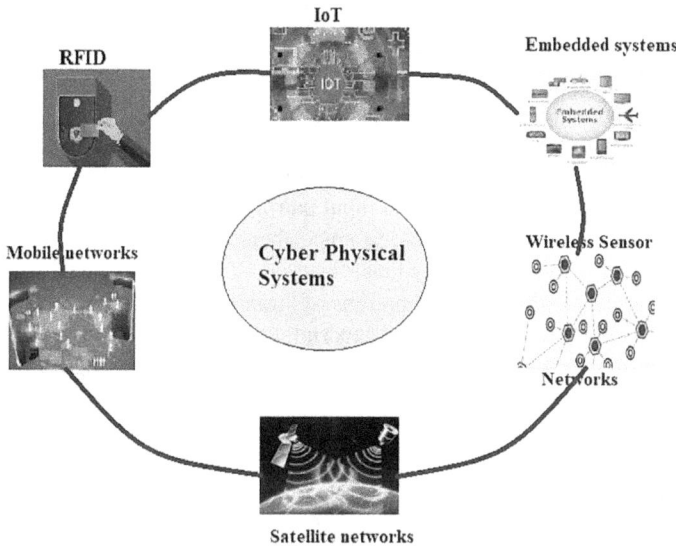

FIGURE 8.1 Components of CPS systems.

The equipment that uses computer algorithms to control and monitor the process is a cyber-physical system. Physical and software elements are closely connected in the network's physical structure. These structures can operate on different space and time scales, exhibit many behaviors, and communicate in different contexts. CPS involves Cybernetics, Mechatronics, Design Philosophy, and Process Science. Integrated systems that are embedded together are often referred to as process control. A combination of components that work together to perform useful functions together is known as developed systems. Therefore, the growth, integration, and continuous management of components are affected by the dynamic interaction and effects of the equipment process. Health network as a modern cyber-physical environment perception system includes embedded devices, IoT technology, and cloud storage [1]. Customer satisfaction is the focus of modern business models. Social responsibility can be earned by withholding profits. It considers all aspects of human culture. It sees the modern company as a social and economic organization, usually responsible to society. In information and communications technology (ICT), context-sensitivity applies but is not limited to the ability to consider the situation of people who may be users or devices.

The effective outcome of a number of different computational intelligence techniques on medicinal data can change the amount of clinical applications by increasing prediction value, detecting the origin and progress of disease, and devising appropriate therapeutic treatments, among several other things. For instance, by building an algorithm on numerous pictures of a specific health condition from a group of previous patients, the occurrence of the same health condition may be recognized in a new patient by feeding the algorithm the newest patient's picture.

A remote care system is comprised of wearable sensors [2] that collect patients' data, transfer it to the data center, and operate it by utilizing machine intelligence to improve prediction accuracy, lower health-related costs by reducing in-patient efficacy [3], and open up new market opportunities [4].

The accuracy of most algorithms is dependent on an initial "learning" phase during which the algorithm understands how to anticipate the output using a collection of training data that contains specified input pairings. This is in contrast to groups of techniques that do not require training since their aim is to identify relationships between the data without explicitly labeling some of the data as "input" or "output" to the computer algorithm [5]. The notions of "training" and "pattern classification" offer two opposing forces in the use of superintelligence in healthcare; while a rise in the number of accessible training data consumes a lot of computing power, a reduction in the number of spare training data results in a reduction in computational accuracy, despite the reduced demand for computer capabilities. Given the high importance placed on algorithmic correctness in healthcare applications, it is fair to assume that forthcoming healthcare systems using intelligent machines will make use of massive datacenters, maybe leased from datacenter operators like Amazon EC2 [6].

Prior to picking a computing technology algorithm, the initial step is often to create a model that specifies the input/output data, including a reasonable level of accuracy. The algorithm to choose is very dependent on the data type, the difficulty of the model, and the objective. Because of verifying and comprehending the model's decision-making process, another consideration is the model's interpretability for humans. As a result, the models should easily demonstrate the correlation between different variables as well as how the variable causes the decision-making procedure.

8.2 CPS IN MEDICAL HEALTHCARE – AN OVERVIEW

CPS views business culture as a socioeconomic enterprise that is frequently socially conscious. Contextual sensitivity refers to the capacity of ICT to take into consideration, but not only, the circumstances of people who may be users or gadgets. Even more obviously, location is an issue in this case. As a result, contextual knowledge is generalized for the situation, which is highly specific for smart phones. Medical cloud-based computing increases, which results in price reductions. Health records are transferred more quickly and securely; cloud hosting optimizes back-end processes and simplifies the creation and maintenance of healthcare apps. With the idealization of phone application systems, a digital health network is instantly established, capable of providing convenient care and authentic connectivity [7]. A cellular data port or a digital cellular data port is a mechanical functionality that is used to interact with a main station in the fields of public transportation, taxi services, private cars, support trucks, commercial fleets, army logistics, fish stocks, cable inventory control, and emergency personnel such as cop cars. The planning and data about the vehicle's responsibilities and operations are presented, including design drawings, infographics, and safety advice. Mobile network terminals include a display screen for displaying information and a keypad that can be accessed for entering data. They may also be connected to a variety of auxiliary devices. In the context of SearchUnified Connections, actual connections include a virtually instantaneous transmission of information between intended recipients over any mobile service with little delay. Though there are two critical concerns in the CPS that must be solved [8], the first priority is the reduction of computation and storage costs associated with mobile terminals [9]. Latter challenge is how to ensure the safety and confidentiality of cyber-physical data [10]. An electrical or electromagnetic device is a set of computer interfaces that display and then convert data into a computer system. The showcase was an early example of a photocopy that made use of a screen. We may use a terminal to send short text instructions to a system to do tasks like directory navigation, file copying, and laying the groundwork for much more advanced automation and technical skills. The term "storage costs" includes cash spent on customer orders or inventory management. Storage costs will be a subclass of keeping manufacturing costs, which will include all expenses. In Mobile Health Systems, surgical instruments with cyber operations are used to connect patients and collect and monitor clinical information [11]. A medical device is any equipment that is intended for medical purposes. Patients profit from the diagnosis and management of diseases, as well as the administration and improvement in the quality of life of clients' medical systems. Substantial risk is inevitable in utilizing a gadget for medicinal reasons, and government regulators in each nation must demonstrate that the equipment is safe and dependable with reasonable certainty before allowing the gadget to be sold. This choice initiates Monitoring Diagnostics, which is utilized to assess the abilities and effectiveness of LCD and CRT screens. These instruments might be self-contained, semi-conductive sensors implanted into the person's body to measure side effects in real time [12]. Individual devices transmit confidential health information included in telemedicine to storage facilities that analyze the data [13].

Currently, numerous cloud computing providers, such as IBM, Google, and Azure, offer accurate healthcare knowledge management [14]. Cloud-based medical systems are advantageous as they provide a diverse selection of medical equipment, worldwide accessibility, and efficient data storage [15]. Nevertheless, the movement of confidential material via an untrusted public cloud creates questions regarding the customer's privacy [16]. As a result, it is critical to offer fine-grained security controls over client records in order to safeguard customer records [17]. Medical CPS (MCPS) raises a number of data security issues due to its multiple benefits [18]. Due to the variety of MCPS and the growing use of online and wireless technology, additional particle sizes and risks have been created [19]. Critical health and patient records may be subject to illegal access in the event of MCPS security breaches [20]. Harmful assaults can result in misinterpretation and ineffective treatment, which can result in the loss of life [21]. As instance, faulty medical equipment, including a cardiac-pacemaker, may cause the gadget to malfunction or totally power-off, resulting in disastrous effects [22]. A few of the primary reasons for weaknesses in MCPS are the devices' and networks' increased interconnectedness, although they are meant to remain separated [23]. Due to the MCPS's heterogeneous and informal nature, security measures are constrained and frequently lack interconnectivity [24]. Apart from the risk of injury and legal liabilities associated with hospital equipment, rear accessibility to the entire network is a possibility [25]. Computer technology enables humans to tackle cybercrime by offering educational capabilities and programming flexibility [26]. Numerous AI approaches inspired by nature are increasingly gaining importance in cybercrime detection and diagnosis [27]. Internet security specialists back up a cybersecurity innovations study by predicting that the economic impact of cybercrime will surpass $6.5 trillion by the end of 2022. In 2021, cybersecurity analysts predicted that a cyberattack will occur every 12 seconds. Spontaneous computation is a branch of study that integrates computers with knowledge from a variety of scientific disciplines, including physicists, biochemistry, ecology, math, and technology. It facilitates the development of novel analytical models such as software, circuitry, and wetware. NIC is multidisciplinary in nature. It enables us to create autonomous computing systems that are capable of responding to their environment and implementing self-tuning, personal management, personal diagnosis, personal configuration, and personal healing [28]. When it comes to data security, artificially intelligent approaches appear to be an extremely promising study topic that improves cyberspace privacy recording [29]. In the healthcare industry, data science and cognition deep learning were also used to minimize operational costs, improve rear operations, and effectively deploy insights in order to create more accurate projections. The phrase advanced analytics refers to artificial intelligence systems that are meant to imitate cognition. Numerous AI techniques are required to create representations for the evolution of human functions on a computing device, including machine learning, deep learning, neural networking, natural language processing, and text analytics. Machine Learning is a type of classification technique that is used to construct models from this data. Cognition computing is a term that refers to computers that educate on a large scale, are helpful, and organically interact with the public. Various models of corporate integration can be simplified. Shipping and availability costs are precise. Integrating the e-commerce ERP enables the ERP to be transmitted throughout the buying process, throughout client travel, and for improved administration via the commercial platforms. Expert systems rely heavily on data generated and sent by IoT gadgets to improve decisions. Due to the variability of CPS, it is impractical to create each gadget in a safe environment, rendering all measures from the devices unusable for teaching the Machine Learning model [30].

As a result, the first stage is to describe the conceptual gadget behavior, depending on how complex it is, in order to generate data for learning the Machine Learning algorithms. Various IoT layers, like transport, perception, and actual outcomes, are likely to be exploited when used in conjunction with a CPS and IoT environment. Facts are pieces of information that are utilized to train a Machine Learning model or machine to anticipate the outcomes that perhaps the system is meant to forecast, or to learn a Machine Learning model or machine. An accent or information marking is added to the data if they are using supervised learning or other combinations of methods that necessitate the use of this approach. Among the things that make up the IoT are personal or physical items that have unique identification, a customizable device, and the capacity to communicate data through a wireless connection. Cyberattacks on an IoT and CPS system paradigm might include, for example, node manipulation, computer virus insertion, Denial of Service (DoS) assaults, phishing, Information Transit assaults (e.g., guy attacks, sniffers), and Navigation Attacks [31].

According to the planned study, a Cyber ML model for protecting medical data on the basis of a cyber-physical system is being developed. For a CPS gadget integrated in a clinical patient monitoring, sophisticated cognitive Machine Learning algorithms are being used for cyberattack and intrusion detection. In order to identify threats, it is necessary to conduct an assessment of a secure setting in order to identify illegal tasks that might cause harm to the organization. In the event that a risk is detected, measures are needed to minimize the hazard in the most effective manner possible before exploiting any weak areas. Client privacy and protection policies should be aligned to make sure that the company is adequately protecting patients' rights to privacy and anonymity. This will also guarantee treatment successfully by preventing interruptions that could have a negative impact on treatment outcome.

8.3 BACKGROUND WORK

Dan Tang and colleagues [32] developed a DDoS assault approach to detect intrusion depending on various characteristics of internet traffic as well as an enhanced AdaBoost method, which they called MF-AdaBoost. To measure elements and pick traffic information from networks, they construct a network feature collector that includes network features. The calculating procedure will extract the most useful information from internet traffic while simultaneously reducing the amount of the data transmitted. The collection of components is being used to select the most appropriate categorization function in order to ensure that the tracking system receives effective training. This technique makes use of the AdaBoost algorithm, which is a classifier used in the Machine Learning area. The results demonstrated that their technology is capable of detecting Distributed Denial of Service (DDoS) assaults.

As proposed by SEO JIN et al. [33], Deep Auto-Encoder and Knowledge Extraction were used to identify the IMPersonation Assault (IMPACT). With the use of a layered autoencoder (SAE), common information C4.8 layer for extracting features, deep learning may organize and function on an energy platform by lowering the amount of operations that need to be processed. The Aegean Wi-Fi Disruption Data is used to train. As a result of the suggested IMPACT results, efficiency was obtained at 98.22%, with 97.6% detection and 1.2% false reports, and the latest benchmark techniques exceeded the normal amount by a significant margin. Another important addition to this research is the evaluation of the AWID database characteristics for the purpose of subsequent IDS development.

For the healthcare CPS, Xu et al. [34] developed the Certificateless-Signature-Scheme (CLS) depending on the NTRU grid, which was afterward used by other researchers. Both of these methods are designed on tiny integer values on NTRU circuits and have proved to be robust against the quantum assault. In addition, security studies and performance tuning indicate that the synchronization and assessment costs of our suggested technique are lesser than those of two existing competing quantum resilience systems, resulting in quantum attack robustness being achieved. But when supercomputers become a reality, strategies for quantum survivability prevention will need to be devised in advance.

Meng et al. [35] developed a confidence-based intrusion detection technique (TBIDA) related to behavioral monitoring, which was then implemented. When the discrepancy between two behavioral patterns in Euclidean could be determined, the reputation of a node can be determined and evaluated. During the assessment, we put our technique to test in a real-world MSN environment by engaging with real-life center staff. When it comes to identifying rogue MSN nodes, experimental data suggest that our technique is far faster than other conventional results. When the difference between two behavioral patterns is defined in Euclidean data points, it is able to derive MSN nodes from them. In collaboration with a functional healthcare facility, they have tested their performance in a genuine MSN setting to see how well they perform. According to the results of our experiments, our technique is more effective than previous comparable processes in detecting rogue MSN nodes.

According to AlZubi [36,37], an AI- powered predictive wellness monitoring system is being investigated (AI-HHMS). This technology enhances the secrecy of individuals' real databases, as well as the integration of medical treatment through different perspectives, by working extremely closely with them. These programs involve specialists, professionals, executives, and staff members who work together to make informed choices more quickly and effectively. Moreover, each event, activity, or organization of IoT utilization by setup should include considerations for data security and quality assurance. The AI-sponsored predictive medical system, which is sponsored by the IoT, improves and minimizes the health hazard dealing with medical data sets collected using wearable applications. The research data suggest that favorable outcomes may be obtained for a variety of performance indicators. The gadget achieves 99.75% precision, a 0.06 error rate, and 98.46% positive states, 98.6% information, and 99.66% accuracy. The technique is connected using the MATLAB software program. Various other computational models related to cyber-physical systems have been implemented by Karthiga et al. in [38–44].

It has been proposed that the Cyber ML model should be used to identify abnormality and assault behavior in medical networks in order to safeguard health information and therefore address these difficulties. Malware, encryption, data thievery, phishing, and risks to insiders are all still prevalent in the medical industry. People are becoming more worried that their protected health information (PHI) has been compromised by treasonous acts including those committed by Anthem and Allscripts. In the field of cybersecurity, there are disagreements: Spyware and ransomware, both of which are cloud-based threats, may be used by malicious hackers to encrypt their devices, servers, or whole networks. Spyware along with ransomware are two topics that have come up recently. There are already too many secure cloud-based storage options for patient data.

8.4   CYBER MACHINE LEARNING FRAMEWORKS WITH INTRUSION DETECTION SYSTEM

CPS are the waves of the future in the field of robotic medicine. They will provide high-quality on-the-go treatment, dependability, and speed in protecting patients throughout the year while maintaining the secrecy of their patients. Integrating physical processes with automated computing processes in real time is the goal of cyber-physical systems. In engineering terms, CPS are lively structures that have something to do with the confluence of routine and transmission processes. Healthcare automation is defined as the controlled functioning of a diagnostic or treatment method by physical or electronic means, with the goal of increasing people's ability to notice, exert effort, and make decisions. It may be defined as follows: Compassionate, ethical, and respectful care should be provided in order for society to be as healthy and affluent as possible. Quality includes, along with clinical quality and safety, the provision of individualized therapy for each and every sole. In order to interact with the physical world, cyber-physical systems must be prepared for unanticipated events and adaptable in the event of a subsystem failure. Medical cyber-physical (MCPS) systems enable the seamless integration of physical and calculative elements into life-crucial, context-known networked devices such as medical devices and ambulatory surgical centers. MCPS is a prospective venue for increasing patient medication efficacy and offering rich-quality medication in the context of contemporary internet-based technologies, such as wireless sensors, linked surgical devices, and mobile care. Medicinal sensory devices and actuators in the MCPS continually watch, assess, and determine a patient's health, and offer therapy in real time based on streaming input from medicinal experts or by mechanized therapies utilizing the healthcare sensory equipment and actuator system. Cyber ML is a model developed in this study for identifying cyber-related attacks on medicinal networks in order to safeguard healthcare information by utilizing CPS and intelligent Machine Learning algorithms.

Intellectual machine learning frameworks may be utilized to simulate the intentions of attackers as well as the events that occur on a network. It is not acceptable to enable a conspicuous or hazardous patient to make a medicinal error (also known as "iatrogenesis"). The disorder is caused by an erroneous or incomplete medical assessment, as well as an accident, an illness, bowel, cancer, or another illness diagnosis or therapy. An information security solution protects information against unwanted access and destruction throughout the life cycle of the information it contains. Integrity protection across all apps and platforms is essential for protecting against data protection, phishing, batch processing, and important performance management. This innovation approach makes it extremely simple for companies and healthcare professionals to communicate secure patient data with one another (PHI). Furthermore, as a result of this plan, patients would be able to obtain smooth, well-organized therapy from clinical staff.

The medical cyber-physical ecosystem is depicted in Figure 8.2. The CPS is composed of two main strategies: a computer system to constantly monitor the natural process; sensory devices and actuators to intervene in the natural process and regulate the science-based approach. It is possible to get numerous benefits from the use of a connection network to connect such modules inside each system, perhaps even in particular circumstances. Virtual surveillance, scalability, reliability, and localization of authority are only some of the benefits. The proposed technique allows for the connection of a large range of clinical systems with cyber systems. So, the network system facilitates the construction of physical systems as well as the unbroken and

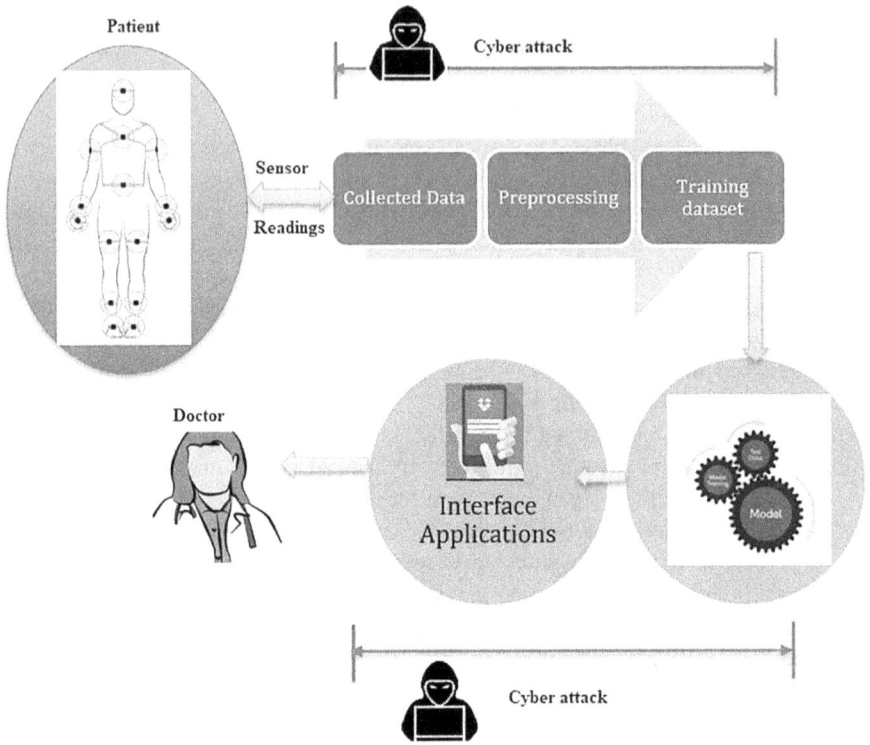

FIGURE 8.2 Cyberattack detection in medical care systems.

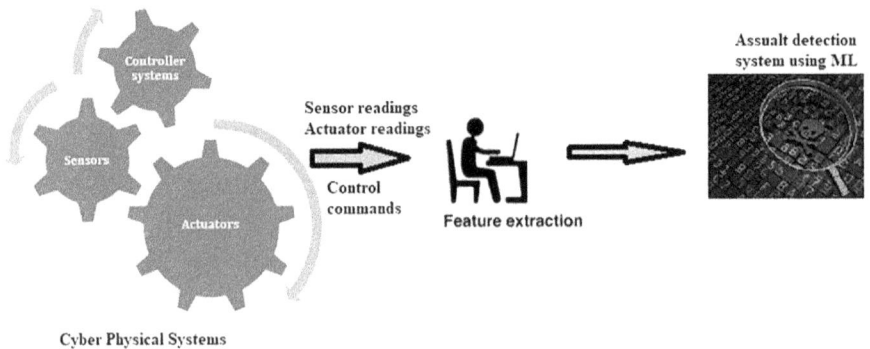

FIGURE 8.3 Proposed cyber ML model.

constant transfer of data for real-world applications. Using the suggested Cyber ML paradigm, the networking costs have been reduced to an absolute minimum.

The discovery of cyberattacks in a smart medical system is depicted in Figure 8.3. An advanced medical system consists of a single or a combination of smart hospital devices (surgically implanted devices, wireless connections, on-the-body devices, etc.) that collect data from a human's body in order to provide better treatment and real-time surveillance to the individual. The Personalized Medical System includes a range of therapeutic (person vitals) and anti-medical (patient physical posture and condition) factors, and it offers significant surveillance to help comprehend the person's condition state in real time. Using an analog signal to collect physiological parameters, smart medical equipment transforms them into a digital data, transfer signals via wireless communication (Zigbee and Bluetooth, etc.) create an electrical gadget such as a computer or desktop or transmit a data packet to a personal smartphone. Using a personalized virtual assistant is similar to using and transferring data from a database (personal servers, public cloud, and so on) to another device. In this case, the database delivers the data to a learning model, which uses AI to select and retrieve features in the data as needed. The Core Information Processing Center employs a system based on Machine Learning. The Core Information Processing Center uses the Machine Learning technique to identify changes in the patient's health, routine patient activities, and SHS risks, among other things. Automatic steps are then performed to enhance patient care (e.g., changing the medication or releasing a new dose of medication), with evaluated data being sent to the authorized organization for further analysis and action. Finally, the doctor provides the patient with a revised personal health monitoring system. Different vital signs are collected by smart healthcare apparatus that is linked to the person's body, for example, electro-encephalogram (EEG), electrocardiogram (ECG), oxygen saturation, etc. Since various technical medical equipment generate a wide range of data sets, attackers might infer that the data produced by the machines is partially dispersed. Accordingly, a new value can be altered within a particular threshold to affect the patient's care or normal routine of activity. An opponent will be aware of the SHS structure, in full or in part, along with the variety of sensors, component correlations, and other information about the design.

An opponent that knows the success labels of the ML model can use this information to launch an assault, such as illness states and normal behavioral patterns. ML systems will be used to identify diseases and user behavior in a variety of situations. An attacker should be aware of the Machine Learning model that is being used to categorize the patient's state. In some ways, Machine Learning-based primary care may be thought of as an information processing network that analyses individuals' vital signs in order to diagnose and cure disease. A medical data collector type collects data across different smart medical equipment that reflects the patient's health status and state and delivers it to a pre-processing data frame for further analysis. Using a data pre-processing paradigm, you may collect data and save it in a domain that is compatible with the relevant sample rates. The selected data is used to train Machine Learning algorithms for real-time surveillance and identification of illnesses, which are then applied to the data. The input images have been labeled with various ailments and innocuous conditions in order to better comprehend the data patterns associated with the various scenarios. The physiological data collected from patients during the testing stage will be evaluated using the previously taught classifier in order to determine various illnesses or medical situations. Our assault approach may be described in light of the data analysis pipeline, which can be found here. An adversarial attempt to implement data collection or data processing in order to alter the initial ML model is described here. A CPS is a computing and information integration of real systems that uses AI. It has the potential to make social interactions increasingly intelligent. Wi-Fi sensor networks (WSNs), which are a key motivating force supporting CPS applications, may be essential for the CPS in the future. The computer chip protection method must preserve firmware from modification, secure recorded data from the gadgets, ensure secure connection, and defend against cyber threats in order to be effective and efficient.

It is possible to do this even by including safety in the design process from the first stages. Storage in the cloud refers to a data center that maintains data on the internet through a cloud-based operator that manages and utilizes information storage as an assistance. It is available on the basis of demand, with a certain amount of capacity and pricing, and it eliminates the need for you to acquire and operate a personalized storage system. It is the study of the human mind and behavior, and its implications for society of technologies, especially VR technology and online networks, is referred to as cyberpsychology. Individuals and groups on the web and in cyberspace are the focus of conventional scientific studies, which are mostly concerned with their psychology. Specialists, employees, patients, as well as other individuals with expertise and private details can benefit from Medical Decision Aid (MDA), which provides medical and wellbeing services that are smartly processed and given at the right moment. MDA is comprised of a number of tools that are designed to support patient procedural outcomes.

Four parts comprise a conventional feedback system: (1) physical events of concern, (2) sensors to detect the physical world and deliver a periodic series x_i representing the physical calculation at interval i, and (3) the microcontroller to deliver a command v_i associated with the input readings that have been received (x_i). There is also a fourth actuator, which converts the command signals into real physical alterations.

A security-related traceability structure for a control framework that operates on the quantum mechanics of the processes necessitates the use of an abnormality prognostication system that obtains as inputs the sensor readings x_i from the real-world devices and the control signals v_i forwarded to the physical devices, and afterward uses them to ascertain whether any suspicious sensors or control commands are present.

8.4.1 REAL-WORLD PREDICTION

Taking the sensor and control signals as an example, a prototype of the real-world systems will identify anticipated future data $\hat{x}_{i+1}$ when the sensor-based and control-based signals are combined. It was decided to use the Auto-Regressive (AR) paradigm for the forecast.

8.4.2 DETECTION OF ANOMALIES

The anomaly identification algorithm is given a time series of residuals r_l (the variation seen between the recorded sensor range x_i and the anticipated measure x_j).

To determine when to raise the alert, it is necessary to conduct tests. Policies for intrusion detection that are long-lasting can be found here. State-full and state-less are two primary categories of intrusion identification policies. Patient-Oriented Design (POD) is a form of User-Oriented Design (UOD) wherein the final is a patient, rather than a healthcare professional who makes use of ICT in medical. Furthermore, to the medicinal consultations and suggestions provided by a medical professional, the provision of specialist care is part of a patient-centric approach. The services are tailored to the beliefs, requirements, and preferences of each individual patient. The medicinal care system is altering its fundamental assumptions. Oriented Design (POD) is a form of UOD in which the final-user is a patient who makes use of ICT in medical care. The medical care framework is altering its fundamental assumptions.

A state-full check is defined as one in which the statistical W_1 is computed and maintained as a watchdog over the previous alterations in r_1, with an alarm being generated when $W_1 \geq \alpha$, that is, when there is a consistent divergence across a large number of timeframes. Many tests may be used to maintain a record of a patient's prior behavior, such as calculating an average over time or using change detection metrics such as the stochastic continuous sum statistics to detect changes in behavior over time. In a state-less check, the alarm is triggered for each substantial variation at a time interval of l: that is, if $\left(x_l - \hat{x}_l \right) = r_j \geq \alpha$, where α represents the threshold.

The suggested Cyber ML model is depicted in Figure 8.3. We have suggested a mechanical technique for detecting attacks in our suggested intrusion framework. It performs cyberattacks at the physical levels of the network by observing and modeling the physical behavior of medical physical systems. Screening and modeling are done through the use of Machine Learning techniques and algorithms. Tracking and modeling concerned with the physical extremes of the cyber-physical system are becoming increasingly complicated. The measurement techniques collected for modeling might be incredibly higher as the cyber-physical system becomes increasingly complex. We put a lot of effort into developing a prominent fingerprint or feature that can be extracted from the noise evaluation used to develop the proposed assault protection system. The features that are produced are excellent at capturing the complex relationships seen between metrics and, more importantly, have higher predictor power in unparalleled basic invasion and procedure events, allowing us to achieve more exact and sturdy predictive accuracy in unparalleled basic attacks and operations.

This chapter focuses on the development of characteristics that may be enhanced to better capture the cyber-physical system resources and are effective in detecting acute attacks from regular things in medical cyber-physical system assault detection systems. Even though attributes can be any signer calculated from any physical measure, the proposed intrusion framework makes use of three different types of attributes: physics-based, learning-based, and survey data. This allows the system components to be seized both temporally and spatially in the custody of the physical systems. This research calculates geographic characteristics based on numerous (multivariate) and separate (univariate) observations, respectively, in the context of spatial computation.

As a prediction model, the Deep Learning Machine (DLM) is employed because of its numerous unique functions. The Deep Learning Machine is a sort of neuro feed network that is unlike any other network. In contrast to the typical feed-forward neural net, in which learning the net entails identifying every link weight and bias, in the Deep Learning Machine, the relationship between concealed and input neurons is randomly chosen and then maintained once the network has been trained. They do not need to be groomed in any way. Because of this, the process of teaching a Deep Learning Machine involves developing the ability to find associations between output and hidden neurons, which would be simply a linear least-square problem, the resolution of which could be methodically addressed by the thorough conversation of the hidden units' output matrix. It is super effective to use the Deep Learning Machine model because of the network's better aim. Furthermore, the Deep Learning Machine outperforms conventional data mining algorithms in terms of reduction efficiency. All regression and classification tasks can be performed using this method.

8.5 RESULTS AND DISCUSSION

The proposed Cyber ML model's experimental outcomes are performed depending on the efficiency metrics like attack determination ratio, attack determination accuracy and delay obtained.

8.5.1 ATTACK DETERMINATION RATIO

Antagonists in Machine Learning-based medical schemes try to modify the distribution of data of the multiple-layer deep learning classifier in order to affect the prediction condition. Certain attacks are focused on medical pictures with the goal of reversing the sickness that has been forecast. When it comes to changing the predicted labels with great confidence, general antagonistic issues may be used to patient data with great success. The suggested approach to discovering vulnerable areas in a clinical time cycle by employing unfavorable assaults on deep prediction models is described in detail. Disease prediction and actual clinical tracking are made possible by Machine Learning-based categorization. The authors of this study believe that the gadgets are in good working order and that the system has zero compromises. Here, a new assault against the enemy will be introduced in order to take full advantage of a smart health service while also altering a patient's condition in order to administer the incorrect therapy. This work provides the placement attack and the evasive assault, which are used to execute the DL model antagonistic attack. The assault determination ratio is depicted in Figure 8.4.

8.5.2 ATTACK DETERMINATION ACCURACY

An assault detection model with maximum accuracy, less connection cost, frequent false positives, and high availability is the major objective of this work, which is also to build an attack detection protection system for use in the MCPS context. Specifically, as seen in Figure 8.5, this work investigates the use of Machine Learning algorithms (Deep Learning Machine) to offer effective attack detection in the HCPS.

Attack Determination Ratios

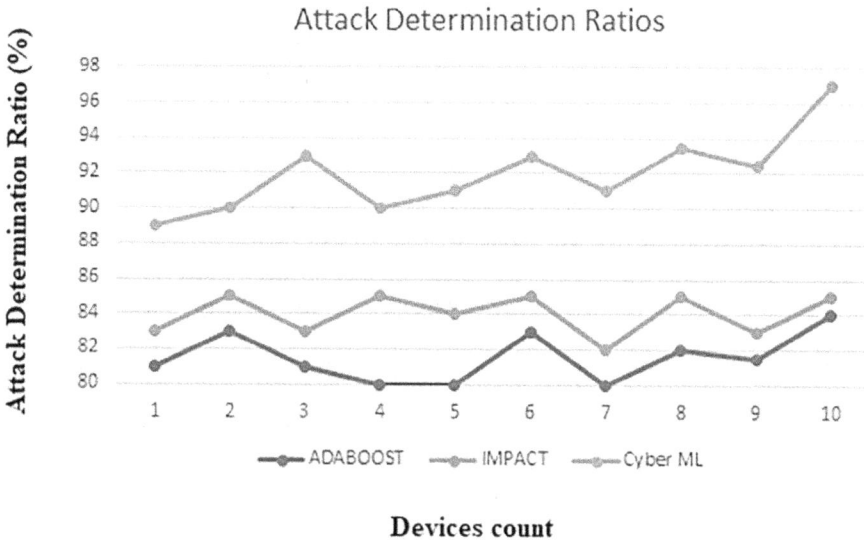

FIGURE 8.4 Attack determination ratio.

ACCURACY

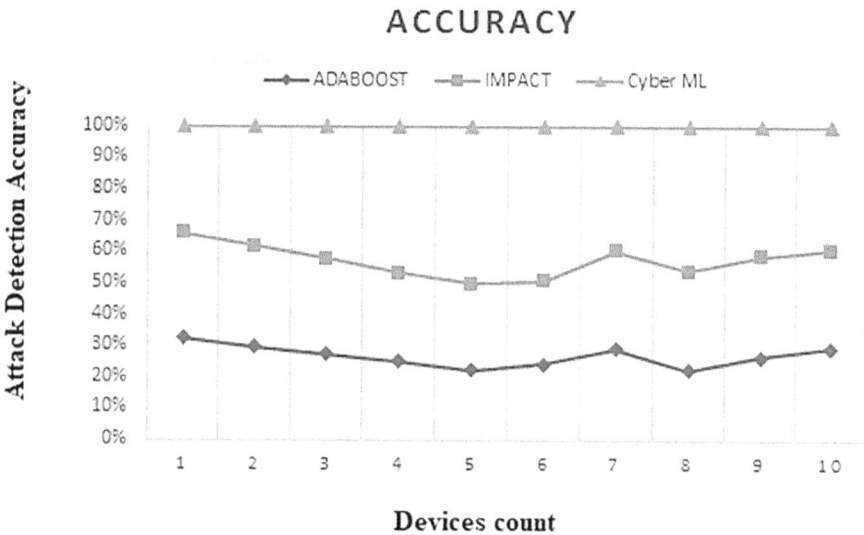

FIGURE 8.5 Attack determination accuracy.

The accuracy of the forecast is determined by the system's performance during the first learning phase, during which it trains how to forecast the outcome depending on a sequence of training data that contains predefined input and output pairings. Also, with special grouping of features, the whole research can result in improved convulsion of the vibrant, nonlinear relationship of the real-world model when obtaining the components with the supervised ML classification algorithm, deep learning console, and good precision in the initial types of fraudulent and attack behaviors when trying to accumulate the components mostly with a deep learning model.

8.5.3   Delay Ratio

System failures are particularly common during specific hours, when the patient's care is transferred to a different caretaker. When a transition is performed with erroneous, imprecise, or unclear data, it raises the likelihood of medical errors occurring throughout the procedure. Confusion related to the delivery of one-way transmission and inaccurate or delayed contact scheduling may be caused by severed cell phone capabilities within a hospital, billing for the inappropriate quality of service, delayed calls, and modification. It is impossible to provide consistent and timely medical care if communication is not established immediately. It can lead to physical issues, long periods of waiting, postponed discharges, bad decisions, and increased tension, to name a few consequences. In order to achieve efficient and dependable care delivery, a network infrastructure that is quick, frictionless, and comprehensive is required. The delayed ratio is depicted in Figure 8.6.

The suggested Cyber ML technique protects patient information while it is being sent across a hospital system. The Impersonation Threat Monitoring Systems (IMPACT) in the Accredited Signature Structure enhanced high attack forecasts, accuracy, efficiency, lowered delays, and connectivity when compared to alternative internet traffic remediation characteristics, and augmented AdaBoost methodologies, according to the observational data.

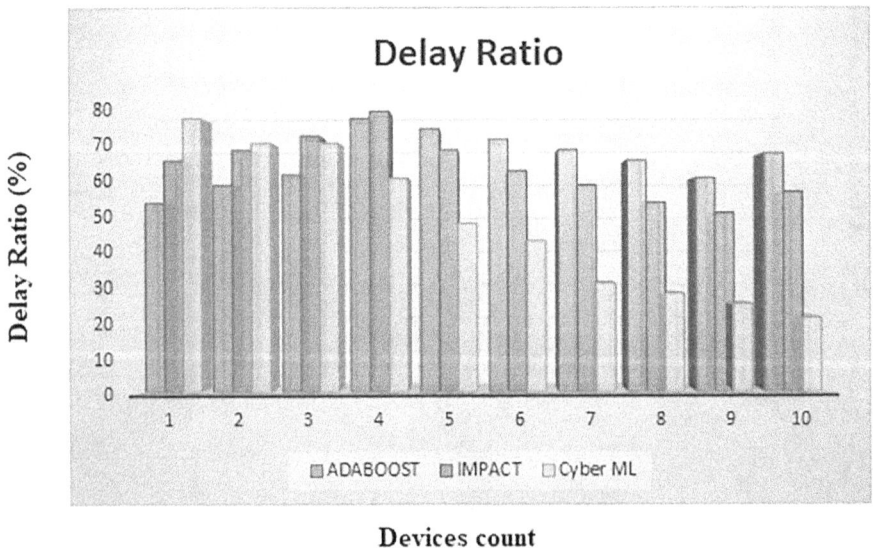

FIGURE 8.6   Delay ratio.

REFERENCES

[1] Shuwandy, M. L., Zaidan, B. B., Zaidan, A. A., Albahri, A. S., Alamoodi, A. H., Albahri, O. S., & Alazab, M. (2020). mHealth authentication approach based 3D touchscreen and microphone sensors for real-time remote healthcare monitoring system: Comprehensive review, open issues and methodological aspects. *Computer Science Review*, 38, 100300.

[2] Hassanalieragh, M., Page, A., Soyata, T., Sharma, G., Aktas, M., Mateos, G., ... Andreescu, S. (2015, June). Health Monitoring and Management Using Internet-of-Things (IoT) Sensing with Cloud-Based Processing: Opportunities and Challenges. In *2015 IEEE International Conference on Services Computing* (pp. 285–292). IEEE.

[3] Kocabas, O., Soyata, T., Couderc, J. P., Aktas, M., Xia, J., & Huang, M. (2013, October). Assessment of Cloud-Based Health Monitoring Using Homomorphic Encryption. In *2013 IEEE 31st International Conference on Computer Design (ICCD)* (pp. 443–446). IEEE.

[4] Page, A., Hijazi, S., Askan, D., Kantarci, B., & Soyata, T. (2016). Research directions in cloud-based decision support systems for health monitoring using internet-of-things driven data acquisition. *International Journal of Service and Computing*, 4(4), 18–34.

[5] Agrawal, R., Imieliński, T., & Swami, A. (1993, June). Mining Association Rules between Sets of Items in Large Databases. In *Proceedings of the 1993 ACM SIGMOD International Conference on Management of Data* (pp. 207–216).

[6] Wankhede, P., Talati, M., & Chinchamalatpure, R. (2020). Comparative study of cloud platforms-Microsoft Azure, Google cloud platform and Amazon EC2. *Journal of Research in Engineering and Applied Sciences*, 5(2), 60–64.

[7] Farivar, F., Haghighi, M. S., Jolfaei, A., & Alazab, M. (2019). Artificial intelligence for detection, estimation, and compensation of malicious attacks in nonlinear cyber-physical systems and industrial IoT. *IEEE Transactions on Industrial Informatics*, 16(4), 2716–2725.

[8] Shakeel, P. M., Baskar, S., Dhulipala, V. S., Mishra, S., & Jaber, M. M. (2018). Maintaining security and privacy in health care system using learning based deep-Q-networks. *Journal of Medical Systems*, 42(10), 1–10.

[9] Hassan, M. U., Rehmani, M. H., & Chen, J. (2019). Differential privacy techniques for cyber physical systems: A survey. *IEEE Communications Surveys & Tutorials*, 22(1), 746–789.

[10] Khan, F., Ahamed, J., Kadry, S., & Ramasamy, L. K. (2020). Detecting malicious URLs using binary classification through ada boost algorithm. *International Journal of Electrical & Computer Engineering (2088-8708)*, 10(1), 997–1005.

[11] Kurde, S., Shimpi, J., Pawar, R., & Tingare, B. (2019). Cyber physical systems (CPS) and design automation for healthcare system: A new Era of cyber computation for healthcare system. *Structure*, 6(12), 1472–1475.

[12] Abdali-Mohammadi, F., Meqdad, M. N., & Kadry, S. (2020). Development of an IoT-based and cloud-based disease prediction and diagnosis system for healthcare using machine learning algorithms. *IAES International Journal of Artificial Intelligence*, 9(4), 766.

[13] Verma, P., Sood, S. K., & Kaur, H. (2020). A fog-cloud based cyber physical system for ulcerative colitis diagnosis and stage classification and management. *Microprocessors and Microsystems*, 72, 102929.

[14] Wang, S., Lei, T., Zhang, L., Hsu, C. H., & Yang, F. (2016). Offloading mobile data traffic for QoS-aware service provision in vehicular cyber-physical systems. *Future Generation Computer Systems*, 61, 118–127.

[15] Weerakkody, S., Ozel, O., Mo, Y., & Sinopoli, B. (2019). Resilient control in cyber-physical systems: Countering uncertainty, constraints, and adversarial behavior. *Foundations and Trends® in Systems and Control*, 7(1–2), 1–252.

[16] Wang, S., Guo, Y., Li, Y., & Hsu, C. H. (2020). Cultural distance for service composition in cyber–physical–social systems. *Future Generation Computer Systems*, 108, 1049–1057.

[17] Sliwa, J. (2019). Assessing complex evolving cyber-physical systems (case study: Smart medical devices). *International Journal of High-Performance Computing and Networking*, 13(3), 294–303.

[18] Wu, D., Zhu, H., Zhu, Y., Chang, V., He, C., Hsu, C. H., ... Huang, Z. (2020). Anomaly detection based on RBM-LSTM neural network for CPS in advanced driver assistance system. *ACM Transactions on Cyber-Physical Systems*, 4(3), 1–17.

[19] Gupta, R., Tanwar, S., Al-Turjman, F., Italiya, P., Nauman, A., & Kim, S. W. (2020). Smart contract privacy protection using AI in cyber-physical systems: Tools, techniques and challenges. *IEEE Access*, 8, 24746–24772.

[20] Iqbal, R., Doctor, F., More, B., Mahmud, S., & Yousuf, U. (2020). Big data analytics and computational intelligence for cyber–physical systems: Recent trends and state of the art applications. *Future Generation Computer Systems*, 105, 766–778.

[21] Poongodi, M., Vijayakumar, V., Al-Turjman, F., Hamdi, M., & Ma, M. (2019). Intrusion prevention system for DDoS attack on VANET with reCAPTCHA controller using information-based metrics. *IEEE Access*, 7, 158481–158491.

[22] Al-Mhiqani, M. N., Ahmad, R., Abidin, Z. Z., Ali, N. S., & Abdulkareem, K. H. (2019). Review of cyber-attacks classifications and threats analysis in cyber-physical systems. *International Journal of Internet Technology and Secured Transactions*, 9(3), 282–298.

[23] Gopalakrishnan, T., Ruby, D., Al-Turjman, F., Gupta, D., Pustokhina, I. V., Pustokhin, D. A., & Shankar, K. (2020). Deep learning enabled data offloading with cyber attack detection model in mobile edge computing systems. *IEEE Access*, 8, 185938–185949.

[24] Qi, L., Chen, Y., Yuan, Y., Fu, S., Zhang, X., & Xu, X. (2020). A QoS-aware virtual machine scheduling method for energy conservation in cloud-based cyber-physical systems. *World Wide Web*, 23(2), 1275–1297.

[25] Haghighi, M. S., Farivar, F., Jolfaei, A., & Tadayon, M. H. (2020). Intelligent robust control for cyber-physical systems of rotary gantry type under denial of service attack. *The Journal of Supercomputing*, 76(4), 3063–3085.

[26] Marques, G., Miranda, N., Kumar Bhoi, A., Garcia-Zapirain, B., Hamrioui, S., & de la Torre Díez, I. (2020). Internet of things and enhanced living environments: Measuring and mapping air quality using cyber-physical systems and mobile computing technologies. *Sensors*, 20(3), 720.

[27] Wazid, M., Reshma Dsouza, P., Das, A. K., Bhat, K.V., Kumar, N., & Rodrigues, J. J. (2019). RAD-EI: A routing attack detection scheme for edge-based internet of things environment. *International Journal of Communication Systems*, 32(15), e4024.

[28] AlZubi, A. A., Al-Maitah, M., & Alarifi, A. (2021). Cyber-attack detection in healthcare using cyber-physical system and machine learning techniques. *Soft Computing*, 25, 12319–12332.

[29] Elhoseny, M., Ramírez-González, G., Abu-Elnasr, O. M., Shawkat, S. A., Arunkumar, N., & Farouk, A. (2018). Secure medical data transmission model for IoT-based healthcare systems. *IEEE Access*, 6, 20596–20608.

[30] Shu, H., Qi, P., Huang, Y., Chen, F., Xie, D., & Sun, L. (2020). An efficient certificateless aggregate signature scheme for blockchain-based medical cyber physical systems. *Sensors*, 20(5), 1521.

[31] Vijayakumar, V., Priyan, M. K., Ushadevi, G., Varatharajan, R., Manogaran, G., & Tarare, P. V. (2019). E-health cloud security using timing enabled proxy re-encryption. *Mobile Networks and Applications*, 24(3), 1034–1045.

[32] Tang, D., Tang, L., Dai, R., Chen, J., Li, X., & Rodrigues, J. J. (2020). MF-Adaboost: LDoS attack detection based on multi-features and improved Adaboost. *Future Generation Computer Systems*, 106, 347–359.

[33] Alarifi, A., AlZubi, A. A., Al-Maitah, M., & Al-Kasasbeh, B. (2019). An optimal sensor placement algorithm (O-SPA) for improving tracking precision of human activity in real-world healthcare systems. *Computer Communications*, 148, 9–16.

[34] Lee, S. J., Yoo, P. D., Asyhari, A. T., Jhi, Y., Chermak, L., Yeun, C. Y., & Taha, K. (2020). IMPACT: Impersonation attack detection via edge computing using deep autoencoder and feature abstraction. *IEEE Access*, 8, 65520–65529.

[35] Xu, Z., He, D., Vijayakumar, P., Choo, K. K. R., & Li, L. (2020). Efficient NTRU lattice-based certificateless signature scheme for medical cyber-physical systems. *Journal of Medical Systems*, 44(5), 1–8.

[36] Meng, W., Li, W., Wang, Y., & Au, M. H. (2020). Detecting insider attacks in medical cyber–physical networks based on behavioral profiling. *Future Generation Computer Systems*, 108, 1258–1266.

[37] Al-Maitah, M., AlZubi, A. A., & Alarifi, A. (2019). An optimal storage utilization technique for IoT devices using sequential machine learning. *Computer Networks*, 152, 98–105.

[38] Karthiga, M., Sountharrajan, S., Nandhini, S. S., & Kumar, B. S. (2020, May). Machine Learning Based Diagnosis of Alzheimer's Disease. In *International Conference on Image Processing and Capsule Networks* (pp. 607–619). Springer, Cham.

[39] Sountharrajan, S., Suganya, E., Karthiga, M., Nandhini, S. S., Vishnupriya, B., & Sathiskumar, B. (2020). On-the-Go Network Establishment of IoT Devices to Meet the Need of Processing Big Data Using Machine Learning Algorithms. In *Business Intelligence for Enterprise Internet of Things* (pp. 151–168). Springer, Cham.

[40] Karthiga, M., Sountharrajan, S., Nandhini, S. S., Suganya, E., & Sankarananth, S. (2021, March). A Deep Learning Approach to Classify the Honeybee Species and Health Identification. In *2021 Seventh International conference on Bio Signals, Images, and Instrumentation (ICBSII)* (pp. 1–7). IEEE.

[41] Karthiga, M., Nandhini, S. S., Tharsanee, R. M., Nivaashini, M., & Soundariya, R. S. (2021). Blockchain for Automotive Security and Privacy with Related Use Cases. In *Transforming Cybersecurity Solutions Using Blockchain* (pp. 185–214). Springer, Singapore.

[42] Karthiga, M., Sankarananth, S., Sountharrajan, S., Kumar, B. S., & Nandhini, S. S. (2021). Challenges and Opportunities of Big Data Integration in Patient-Centric Healthcare Analytics Using Mobile Networks. In *Demystifying Big Data, Machine Learning, and Deep Learning for Healthcare Analytics* (pp. 85–108). Academic Press.

[43] Suganya, E., Sountharrajan, S., Shandilya, S. K., & Karthiga, M. (2019). IoT in Agriculture Investigation on Plant Diseases and Nutrient Level Using Image Analysis Techniques. In *Internet of Things in Biomedical Engineering* (pp. 117–130). Academic Press.

[44] Sountharrajan, S., Nivashini, M., Shandilya, S. K., Suganya, E., Banu, A. B., & Karthiga, M. (2020). Dynamic Recognition of Phishing URLS Using Deep Learning Techniques. In *Advances in Cyber Security Analytics and Decision Systems* (pp. 27–56). Springer, Cham.

9 IoT-Based Smart Irrigation and Monitoring System for Agriculture

K. P. Sampoornam and S. Saranya

Bannari Amman Institute of Technology, Erode, India

Hyder Ali Segu Mohamed

College of Engineering and Information Technology, Buraydah, Kingdom of Saudi Arabia

CONTENTS

DOI: 10.1201/9781003181668-9

9.1 INTRODUCTION

Agriculture is our country's number one food production source. In India, agriculture contributes significantly to 18% Gross Domestic Product (GDP) of the country, more than the half of the population is employed by this, where economy is primarily based on the agriculture and isotropic weather conditions, yet it is unable to make the use of the agricultural resources. Major reason for this is scarcity of rain and groundwater. Agriculture [1] is the world's largest freshwater consumer, accounting for up to 70% of the total use, which advocates smart management of water to ensure food and water safety for the population of the world. The water irrigation system and methods of crop cultivation field application play an important part in this. To avoid productivity loss due to under irrigation, farmers splash as much water as is needed that is called overirrigation, and which results not only productivity challenges, but there is also wastage of energy and water.

The continuous demand for food requires quick enhancement in technology for food production. A need for smart farming [2] has grown to a greater extent particularly in developing countries such as India. Besides research has taken on a new dimension in the wireless sensor network which is based on Zigbee in agriculture like monitoring environmental conditions such as soil moisture content, weather, crop growth, and temperature.

Nowadays the awareness of implementing technology for the agricultural environment has increased. Manual data collection can be sporadic, not continuous for desired factors, and can generate differences from the incorrect measurement. This can provoke the controlling of the main environmental factors to be difficult. Separate, wireless sensor nodes could even reduce the effort required to monitor the environment. Data logging allows the lost or misplaced data to be reduced. It would also allow personnel to be placed in critical places without having to put them in dangerous situations. Surveillance systems can ensure faster response time to harmful conditions and influences.

IoT [3] arises as a natural option for smart management applications of water; although, the implementation of the various technologies needed to make it function smoothly is still not completely achieved in practice. The development of IoT is a phenomenon due to the combination of many influences like low-power wireless technologies, low-cost devices, the cloud data center's availability for processing and storage, high-performance commodity platform computing resources, frameworks management for dealing with unorganized social network data, and a computational intelligence algorithm for dealing with this data of monumental amount.

Smart farming [4] enables farmers to produce good yields using minimal resources like fertilizers, water, pesticides, sunlight, etc. Farmers can install the sensors to know their resource crops and decrease environmental circumstances impacts. Smart farming [5] is also referred to as precision farming because numerous sensing devices are used to provide data that helps the farmers to optimize and monitor the crops and also become accustomed to factor changes in the environment.

9.1.1 FARMING IN INDIA

About 2500 years ago, the Indian farmer found and started cultivating many spices and sugarcane. Farming actually contributes as much as 6.1% to our GDP as of 2017. One of the oldest economic operations in our country is farming. Different areas have numerous agricultural practices. However, with improvements in weather and climate patterns, technical advances, and socio-cultural traditions, all these approaches have greatly changed over the years. It is possible to describe farming methods prevalent in India as follows.

It is a primitive form of farming, and it is still practiced in some areas of the world by farmers. While this form of subsistence farming is usually carried out on small land fields, it often uses traditional instruments such as a Dao, hoe, digging sticks, etc. Typically, this farming system entails a family or local group of Indian farmers who use the produce for their own consumption. This is the most natural technique in which crop growth is based on weather, sun, soil fertility, and other environmental conditions.

The "Slash and Burn" method is the secret to this farming technique. In this activity, farmers burn the field until the crops are cultivated and harvested. For a new batch of planting, they then transfer to a clear patch of land. As a result, the soil, inevitably, regains its fertility. Since no fertilizers are used for agriculture, the primitive system of survival yields high quality crops and preserves the soil's properties as well.

A. Subsistence Farming
   This is farming that can be either primitive or intensive, and is performed for the farm owners' use. The only goal here is to fulfill the desires of the farmer and his kin. Primitive subsistence agriculture is the type of subsistence farming that is traditionally performed using traditional tools such as dao, hoe, digging sticks, etc., on small areas of land. This is the most sustainable method of growing crops, since rain, sun, wind, and soil conditions lead to the growth of crops in the natural world.
B. Commercial Farming
   It becomes commercial agriculture as farmers cultivate crops and rear animals for economic activity. Farmers cultivate greater fields of land, with heavy use of machines, because of the need for a high volume of production.

C. Home Farming

Home cultivation involves farming on the terrace and gardening. Limited space and small tools including a garden rake, pruning shear, etc. is needed. In the same land, this farming has the potential to grow some vegetables, fruits, flowers, and small trees. This farm is often used as a decorated household piece. Little labor is required. This agriculture is used as both industrial and subsistence agriculture.

9.1.2 SMART FARMING

Smart farming relies on the use of information collected in the management of farm operations from multiple sources (historical, geographical, and instrumental). Technologically sophisticated does not really mean that it's a clever machine of nature. By their ability to track the information and make meaning out of it, smart systems distinguish themselves. Smart farming uses hardware (IoT) and software (SaaS) to collect data and offer actionable information to handle both pre- and post-harvest activities on the farm. The documentation on any part of finance and field operations which can be tracked from everywhere in the world is structured, available all the time and full of data.

The UN predicts that by 2050 the world's population would reach 9.7 billion, triggering a 69% growth in global agricultural demand between 2010 and 2050. Farmers and agricultural firms are moving to the IoT for analytics and higher development capability to satisfy this need.

Agricultural technical advancement is not new. Hundreds of years ago, manual devices were the standard, and then the industrialization brought in the cotton gin. The 1800s resulted in chemical fertilizers, grain elevators, and the first tractor powered by coal. Quick forward to the late 1900s, when farmers started preparing their work based on the satellite information.

The IoT is expected to drive agriculture's future to the next level. Smart farming has already become more popular with farmers, and thanks to agricultural drones and sensors, high tech farms are increasingly becoming the norm. IoT technologies in agriculture and how "Internet of Things farming" will assist farmers in the coming years to satisfy the world food demand are discussed in the subsequent sections.

9.1.2.1 Precision Farming

In order to enhance the efficacy of their day-to-day jobs, farmers have also begun employing certain high-tech farming methods and innovations. Sensors installed in fields, for example, allow customers to produce accurate maps of both area's topography and resources, and also variables like temperature and soil acidity. To predict weather conditions in the coming days and weeks, they can even access climate forecasts.

Farmers can use their smartphones to track their crops, livestock remotely, and equipment, as well as collect stats about feeding and producing their livestock. They can also use this technology for their crops and livestock to run mathematical forecasts. And drones are becoming an invaluable instrument for farmers to survey their fields, conduct field research, and produce data in real time.

John Deere (one of major brands in agricultural machinery) has started linking its tractors to an internet as a concrete example but has developed a method to show data on the crop yields of farmers. The business is pioneering self-driving tractors, similar to smart vehicles, which will free farmers to perform other activities and improve productivity further.

Both of these methods lead to precision farming, the practice of using satellite imaging and other tools (like sensors) to observe and record information in order to increase production performance while reducing costs and maintaining energy.

9.1.2.2 Smart Greenhouse

In order to build a self-regulating microclimate conducive to crop production, smart greenhouses exploit IoT and linked devices. Such managed environments remove the challenges of bad weather and pests while supplying farmers with real-time insights for optimal production. In order to manage irrigation, crop spraying, illumination, humidity, temperature, and much more, farmers utilizing smart greenhouse crop monitoring systems will exploit information from big data and analytics.

9.1.2.3 Future of Farming

Smart agriculture and precision farming are beginning to take off, but they may only be the precursors to the world of agriculture's much greater use of technology. The growth of block chain technology is finding its way to IoT, and because of its potential it can provide valuable crop data, it may be essential in the farming field. Farmers can use sensors to capture crop information entered on the block chain that includes recognition variables, salt and sugar content, and pH levels.

There will be almost 12 million agricultural sensors installed worldwide by 2023 for Insider Intelligence projects. In addition, the tech giant IBM estimates that half a million data points per day can be generated by the average farm, assisting farmers to develop yields and maximize wages.

It is reasonable that farmers are gradually looking to agricultural drones and satellites for the future of farming, considering all the possible advantages of these IoT applications in agriculture.

Drones enable farmers to track how much further crops are in their respective periods of growth. In addition, to bring them back to life, farmers may spray ailing crops with substances using drones. Drone fly reports that drones inject manure 40–60 times quicker than doing so by hand.

9.1.3 Differences between Traditional and Smart Farming

Traditional Farming	Smart Farming
1. Manual management of both sector and finance data independently, resulting in mistakes	1. Early identification and deployment only in the affected area, cost savings
2. Same range of cultivation methods throughout the area for a crop	2. Each farm is evaluated to see the necessary crops and water needs for efficiency.
3. No way to make weather predictions	3. Analysis and forecast of the weather
4. Not practical for geo-tagging and zone detection	4. In farms, satellite imaging detects the numerous zones
5. Fertilizer and pesticide use in the field	5. In the same area, field and finance data are available showing profiles, yields and trends with clear reports.

9.2 LITERATURE SURVEY

Author Boman B. et al. [6] employed open-loop systems where the controller chooses the water quantity to be used and an irrigation event is scheduled. Correspondingly, the water is applied as per the schedule requirement by programming the controller. The disadvantage of the open-loop system is the incapability to automatically respond to environmental condition changes. Additionally, they may need regular resetting to reach optimal irrigation efficiency levels.

Sarode K. R. et al. [2] proposed the microcontroller-based system for controlling various agricultural field parameters. For long-distance communication, the proposed system uses ZigBee technology. This system is expected to help farmers assess soil and weather by crop. The device is built for farmers to be better off. Applications of the smart sensor-based agricultural monitoring device have been used to raise crop yields by monitoring environmental factors and supplying information to observers. In the current scenario with uncertain parameters, it will be a promising technology for farmers all over the world.

Nikesh Gondchawar et al. [7] aimed to offer smart farming based on IoT. Remote controlled robots based on smart GPS, smart warehouse management, and smart irrigation are the methods for smart agriculture. These all methods are managed by an internet-connected smart device or the computer and operations of all these methods are carried out with the aid of sensors. Node 1 in architecture of this network is a remote-controlled robot based on smart GPS that performs tasks like spraying, moisture detection, scaring birds and animals, maintaining vigilance, weeding, etc. The smart warehouse management node 2 has been used to monitor temperature and humidity by sensors. Smart irrigation node 3 is used to manage level of the water pump mechanism until moisture reaches to the soil. AVR Microcontroller Atmega 16/32, Zigbee board, raspberry pi, moisture sensors, temperature sensorLM35, and dip trace, AVR studio version 4, raspbian operating system, proteus 8 simulator, sinaprog, are the hardware sensors used to perform the operations. All of this IoT automation is used for the purpose of growing crop yield and general production.

Author Anitha K. [8] has developed an automated irrigation system which, depending on soil moisture, turns the ON or OFF motor pump. The use of the proper method of irrigation is important in field of agriculture. In the proposed system, 8051 series microcontroller is used, which is coded to receive the input signal of the differing soil moisture conditions via sensing arrangement. Sensing configurations are done using two stiff metal rods that are inserted at a distance into the field. The control unit is interfaced with the metallic rod connections.

Akyildiz, I. F et al. [9] aimed to grow the high yield crop in the proposed system and to minimize human effort. Some sensors are used in the proposed system to measure the soil of the seed. In agricultural field, the sensors like temperature sensors, humidity sensors, pressure sensors, and the humidity sensors are used. The temperature difference in the shape of the land is used to increase nutrient content in the shape of the ground. The moisture sensor acts on an electrical conductivity principle. One of the essential considerations for crop growth is the moisture content. To manage the flow of water, pressure sensor is attached to the microcontroller. The humidity sensor is used to determine the level of humidity in the air.

Garcia-Sanchez, A. J. et al. [10] used Node 1 as the mobile robot sensor in the proposed system. This is used to remotely power water pumps. Water content level is less when pump is automatically ON or the water content level in the soil is more when the pump is automatically OFF. In Node 2, in raspberry pi, sensors such as motion detector, temperature sensor, humidity sensor, light sensor, space heater were used. Temperature sensor was used to determine temperature level of the farmland. The moisture sensor in Node 3 is used to test the soil quality of agricultural fields. The data that is transmitted is sent to Node 2 and is sent to the microcontroller. To monitor the water pumps, the data is used. The Raspberry pi is a compact device used for networking and computation. All data is submitted to cell phone farmers. Information is sent via GPS to the base station. The data is sent by microcontroller to Raspberry Pi.

Carlos Kamienski et al. [11] introduced SWAMP design, platform, and device implementations that illustrate the platform's reliability, and scalability is a key problem for the IoT applications, the performance analysis of the FIWARE modules used in the platform is presented. Results suggest that the platform is capable of providing the SWAMP pilots with adequate performance, but needs specifically made configurations and re-engineering of some modules to provide the minimum computing resources with improved scalability. The SWAMP project implements an IoT-powered smart management of water network with such a hands-on approach based on four pilots in Europe and Brazil for accurate irrigation in agriculture.

Sushanth et al. [12] developed the system that can monitor humidity, moisture, temperature and even the movements of the animals in agriculture field that destroy the crops in the field by using sensors and Arduino board as a controller. The SMS notification is send to the farmer's smartphone in case of any emergency by using the Wi-Fi. The device has duplex communication connection based on the cellular internet interface which enables for irrigation scheduling and data inspection to be configured via an android platform. The device is having the ability to be effective in limited water due to its electricity sovereignty and low price, geographically isolated regions.

Nogueira L. C. et al. [13] used open loop models wherein the operator decides on the quantity of water to be used and the duration of irrigation. Controller is configured accordingly and water is applied according to appropriate timetable. For control purposes, open loop control systems use either the irrigation length or the specified applied amount. Normally, a clock which is used to begin irrigation comes with open loop controls. Irrigation is terminated on the basis of pre-set time or on the basis of a defined amount of water that passes through the flow meter. The user takes the decision in an open loop system about the volume of the water which is added and when irrigation incident occurs. In the controller, this data is configured and water is added according to the desired timetable. The difficulty of open loop design is the inability to adapt to varying environmental factors automatically. Moreover, to attain high levels of irrigation quality, they can require regular resetting.

A remote sensing and irrigation control system using a distributed wireless sensor network aimed at variable rate irrigation, field sensing in real time, and control of a site-specific precision linear movement irrigation system. Maximizing utility with minimal use of water was defined by Y. Kim Kim [14]. The device defined specifics of the variable rate irrigation architecture and instrumentation, the wireless sensor network, and actual field sensing and control through the use of suitable software. Five field sensor stations were used to create the entire device, which collects the data and sends it to the base station using the Global Positioning System (GPS) where appropriate irrigation management steps have been taken in compliance with the database available with the system. The scheme offers a method as well as a remote control for precision irrigation, the promising low-cost wireless solution.

9.3 METHODS OF IRRIGATION

Irrigation was always a traditional procedure that has emerged over the years, through several stages. Our ancestral farm workers sought various methodologies to irrigate their farm . Manual irrigation with drip irrigation, watering cans and buckets, irrigation with sprinklers, and flood irrigation which have been popular and are still used today. There are several limitations to the existing system: soil nutrient leaching, flood erosion, water loss through evaporation from plant surfaces, water wastage that may lead to drought area's water scarcity, and unhealthy crop production. Following the technological developments of the past three decades, channel irrigation, sprinkler irrigation, and Drip Irrigation are the three major irrigation methods used to drench the agricultural land, and these three class of irrigation methods varied according to crop requirements.

9.3.1 CHANNEL IRRIGATION

Channel irrigation is a type of surface irrigation where water is applied to the crop via channel network. It is the most significant form of irrigation mostly in developing countries in several parts of the world. It is cheaper and has the greatest advantage in the regions of the river valley. The irrigation of the canals in India is one of the main methods used to improve crop growth. In India, U.P. did the first use of the canal irrigation, then Haryana and Punjab. However, this method is extended only to those areas that have deep fertile soil large-level plains drained by well-distributed river basins. Irrigation of the canal is of great use in the deltas of rivers such as Krishna, Godavari, Kaveri, Ganga, and Mahanadi in the coastal plains of Kerala. Figure 9.1 shows the channel irrigation.

The main drawback of this irrigation method is that water will be lost either by surface evaporation or by flow through the channel peripheries. The water lost by evaporation is generally small in comparison with the water lost by the inlet. The loss of evaporation is generally 2–3% of the total losses, but it may exceed 7% in the summer.

9.3.2 SPRINKLER IRRIGATION

Water is circulated from the central location in field or from the sprinklers on moving platforms via overhead sprinklers or guns with high pressure. The water is pumped to one or more prime locations within field in a sprinkler or overhead irrigation and circulated by sprinklers or guns with high pressure. A system that uses overhead mounted guns, or sprayers, sprinklers on the installed risers permanently is very

FIGURE 9.1   Channel irrigation.

FIGURE 9.2   Sprinkler irrigation.

usually referred to as an irrigation device with a strong collection. Rotors are called larger pressure rotating sprinklers and are driven by the gear drive, impact mechanism, or ball drive. Rotors may be intended to rotate in the partial or full circle. Although this lowers the manual effects, water wastage will be more compared to the method of drip irrigation. Sprinkler irrigation is shown in Figure 9.2.

9.3.3 DRIP IRRIGATION

In drip irrigation, farmers can save more water as it supplies the water directly to the plant's root on the soil's surface in the form of droplets. Water is delivered, drop by drop, to or near the plants' root zone. If correctly handled, this system would be the most water-efficient irrigation method, reducing runoff and evaporation. When properly managed, in the drip irrigation; field water quality is usually within the scale of 80–90%. Figure 9.3 shows drip irrigation.

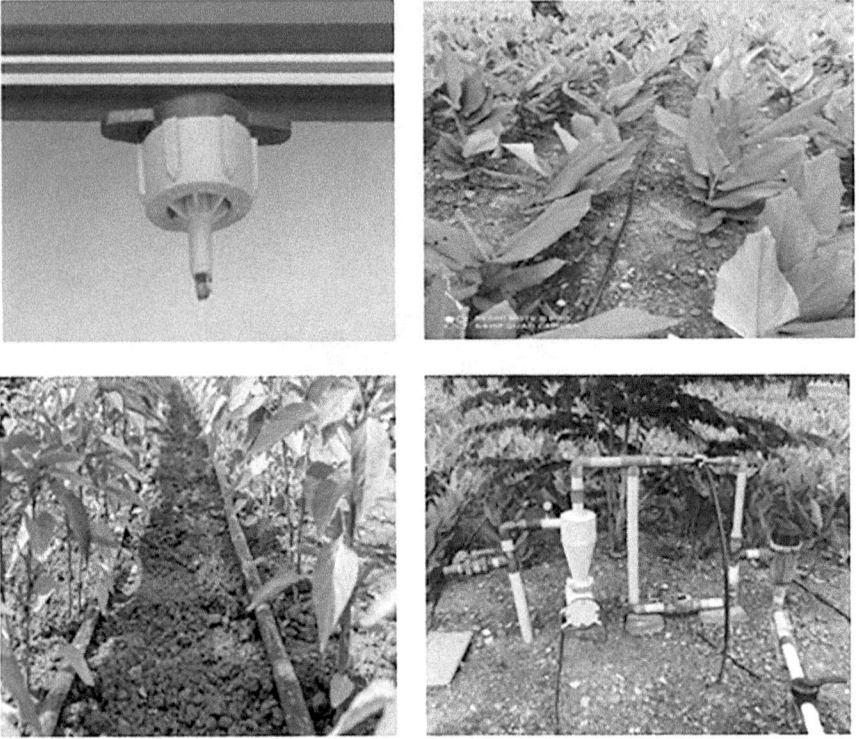

FIGURE 9.3 Drip irrigation.

9.4 COMPARATIVE STUDY

In Table 9.1, five different agricultural systems are compared based on the technologies used and advantages of the system in agriculture field. Only the smart irrigation system provides data acquisition and cloud implementation.

9.5 WSN-BASED IoT SYSTEM

9.5.1 WSNs

Wireless sensor networks (WSNs) in the wireless communication-based environment consist of several sensor nodes. The sensor node is intended to track physical processes with minimal resources and memory, such as temperature and humidity [20]. The focus of many recent researchers is on using various strategies to overcome these drawbacks, like various routing protocols, intelligent methods, etc. In different settings, they have many uses, like military, medical, schooling, agriculture, control systems, etc. By improving the IoT model, it is expected that further workspaces will appear. Figure 9.4 shows the prototype architecture of a node in the sensor.

The node configuration consists of modules for the sensor, processor, communication, and control. The A/D converter was used in the sensor module to transform

TABLE 9.1

Comparison of the Different Agricultural Systems

Existing System	Year	Technologies Used	Cloud Implementation and Data Acquisition	Advantages
Cucumber disease detection [15]	2017	Artificial neural network, MATLAB.	Only data acquisition	High accuracy
Detection and Identification of disease stages [16]	2014	ASD spectroradiometer, MATLAB.	Only data acquisition	This SDI exhibited high accuracy and sensitiveness.
Smart irrigation System [17]	2015	MATLAB, IOT, wireless sensor,	Both cloud and data acquisition	Water usage optimization, control the system
Identifying and monitoring winter wheat diseases [18]	2015	Hyperspectrum	Only data acquisition	Diseases are determined and differentiated.
Non-linear analysis of soil microwave heating [19]	2017	Microwave antennas, electromagnet ethic heating.	Only data acquisition	Provides an effective solution

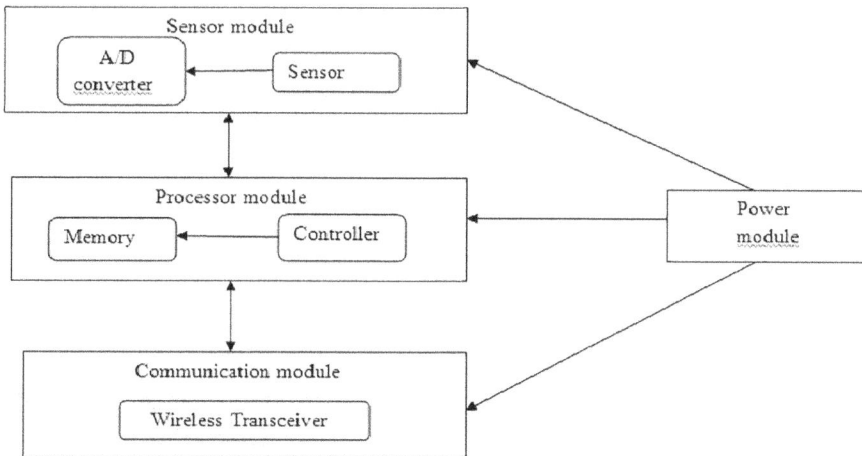

FIGURE 9.4 Node architecture.

the analog signal into a digital signal that is said to be a sensor output. The key function of processor module is for instructions to be executed. The processor's memory retains the internal clock and sensed data briefly. The communication module's wireless transceiver consists of the both transmitters and receivers that share a shared circuit. In Module Power, battery is used with specified watts of absolute power [21].

9.5.1.1   Function of Nodes

WSN node has the capability to sense the world and transmit information obtained from the visible area. As nodes gather information, they compact it and relay it directly or indirectly through BS with the aid of the other nodes. Transmitted data is provided to the device with the aid of the gateway connection. The gateway nodes serve as an interface between the nodes and the server where the data gathered from all the nodes is processed. When the node is unable to connect directly with any other nodes, it requires other nodes to forward data. Multi-hop routing is said to be this method of network data transfer. The transmitted data may be used inside area (locally) or around area (globally). A node consists of less power sensing equipment, power module, embedded processor, and communication channel.

The nodes feel the phenomenon's actual occurrence and the sensing device translates them to digital signals. Then they do, and then they use the control machine to manage captured data and process it. This machine is comprised of subunits for processing and memory. In their database, necessary data can be stored. Furthermore, they do support the requisite operations by means of a control unit from the few hours to months or years. Often, sensor nodes are capable of communicating centralized server/base station or together decentralized designs focused on multiple topologies, such as Mashes, stars, etc. Via the contact machine, this is understood. The sensor nodes can interact point-to-point or multi-hop models.

For classic sensor nodes, these are the deepest differences. In addition, these nodes are commonly distributed in an environment from 1 meter to 100 meters. These nodes can be used in low bandwidth communication with restricted processing capacity.

The features of the wireless sensor nodes are usability, wide-scale implementation, network mobility, and resilience. The lack of single node would not affect the overall performance of network, since these networks are involved in physical information rather than information from a single sensor. Fault tolerances are, however, a significant design element in networks. There is a need to be mindful of other causes. These networks are inexpensive. They are also practically used as the low-cost sensor nodes. New sensor systems with various hardware and software characteristics are being introduced by researchers.

9.5.2   Internet of Things

The IoT is one of the recently explored technology in which fields of use are increasing rapidly. Most of the nodes that are fitted with the output of internet are composed of IoT technique. The new paradigm is based on emerging developments such as ad hoc networks, embedded devices, wearable technology, and creation of IoT and machine learning techniques. Link to the internet problem is one of the most critical specifications of this technology. The internet is one of the inseparable aspects that can minimize costs and time in our lives. Imagine making internet surf for the watch that missed in the house somewhere [22].

This is therefore the core vision of the IoT, a world in which things can speak and process their data to execute desired tasks via methods of deep learning. IoT is

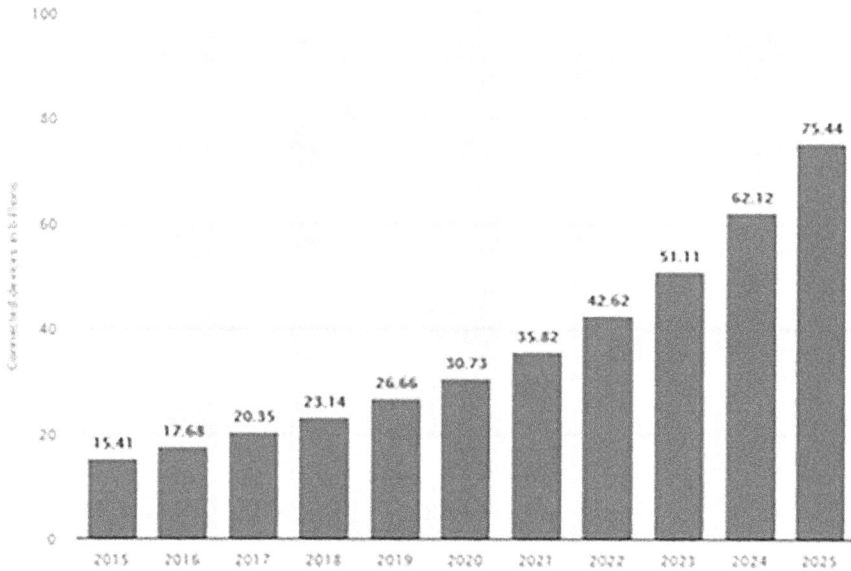

FIGURE 9.5 IoT-connected devices.

designed around the wireless radio waves which allow various devices to communicate through the internet with each other. Any protocols like less-power Bluetooth, Wi-Fi, RFID, NFC, and so on are included in this network. The physical object in the IoT gathers and processes the data from the atmosphere and/or other objects that it can obtain [23]. Sensors, actuators, internet access, and network communication are embedded with these objects. In reality, the connected internet does not have to be all computers. The cost of the machine can be lifted by this house. In this case, the WSN structure may be used by a device builder to access a smart program.

Examples of IoT implementations are environmental and personal health management, monitoring and regulation of manufacturing systems, such as irrigation, smart spaces, and smart cities. According to Figure 9.5, nearly 75.44 billion computers are expected to be wired to the internet by 2025. The artifacts interact together in IoT-based agricultural applications to provide valuable knowledge from the field or greenhouse . In other words, for reducing the participation of the human element, increasing efficiency, as well as reducing costs, the internet and physical agents have been successful. Figure 9.6 shows wireless sensor network-based IoT system.

9.6 HARDWARE DESCRIPTION

Sensors are widely used to detect parameters and respond to any of the electronic controllers for information acquisition and collection. Sensor-gathered information is usually a data signal and is converted to a human-readable format and is displayed on a display. Within this system, the following types of the sensors are used.

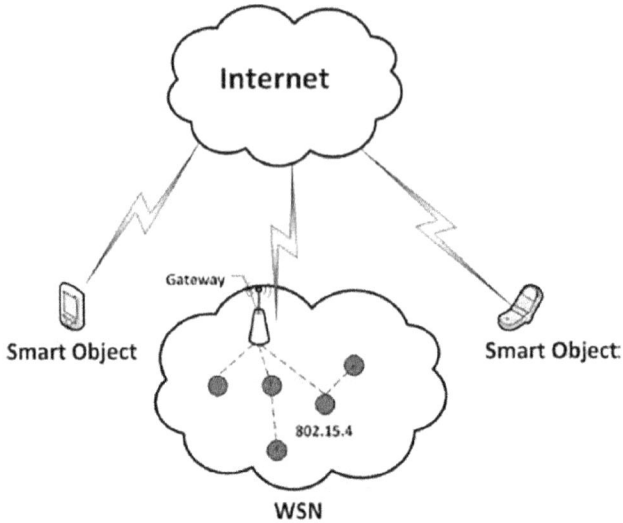

FIGURE 9.6 WSN-based IoT.

9.6.1 Soil Moisture Sensor

Figure 9.7 shows the soil moisture sensor. This sensor is used for assessing the volume unit content of the soil water level. This sensor is capable of measuring its value because of the plant uptake and evaporation which leads to the loss of moisture. It can also analyze the soil moisture content desired for different crop species. This sensor can observe soil moisture in the range of 10 meters above ground level. It exploits the properties of the soil's electrical resistance. The relationship between the measured property and soil moisture is calculated and can differ based on the property being measured. Factors in the environment, such as temperature, form of soil, or electrical conductivity plays a major role. Here, the moisture in the field is sensed and passed to the microcontroller to monitor the switching operation of the water pump when ON/OFF is done.

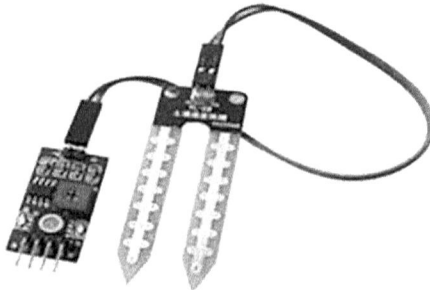

FIGURE 9.7 Soil moisture sensor.

9.6.2   LDR Sensor

A cadmium sulfide (CdS) or a photoresistor cell is also called a Light-Dependent Resistor (LDR). This is also known as photoconductor. It is photocell; the principle used for this is photoconductivity. The resistor is a basic passive component, whose resistance value decreases with decreased light intensity. LDR sensors shown in Figure 9.8 are used to detect light. The resistive LDR sensor ranges from 400 Ω to 400 kΩ.

9.6.2.1   Working of LDR

Photoconductivity, which is nothing but an optical illusion, is the operating theory of the LDR. If the substance absorbs light, then the material's conductivity decreases. If the light lands on the LDR, then the electrons in the material's valence band are ready for the conduction band. But to allow the electrons jump from one band to another band, the photons in the incident light must have energy superior to the element bandgap (valance to conduction).

Therefore, as light has enough energy, the conduction band that scores in a significant number of charge carriers is stimulated by more electrons. The resistance of the system reduces as the influence of this approach and the flow of the current continues to flow more. The working of LDR is shown in Figure 9.9.

9.6.3   PH Sensor

According to its pH value, soil is alkaline, neutral, and acid. Many plants require a pH range of 5.5–7.5, but few species require more alkaline or acid soils. Nevertheless, for optimum growth, each plant requires a specific range of pH. pH has a strong influence on nutrient availability and the presence of plants and microorganisms in the soil. For instance, fungi require acidic conditions whereas most of the bacteria prefer moderately acidic or slightly alkaline soils, especially those which provide plants with nutrients. To optimize its growth, each plant needs elements in different

FIGURE 9.8   LDR sensor.

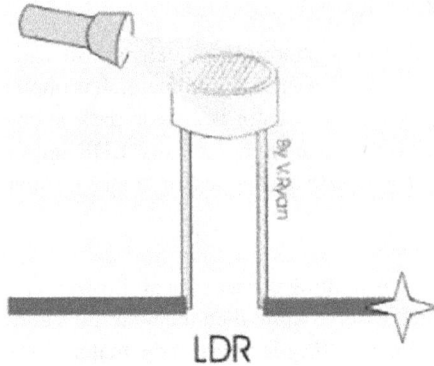

FIGURE 9.9 Working of LDR.

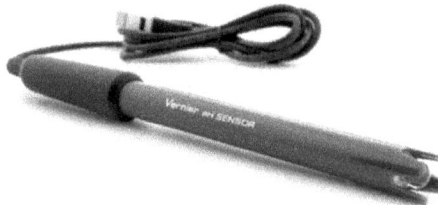

FIGURE 9.10 pH sensor.

amounts of pH. pH sensor determines the level of soil acidity. It can be used for a long term. To determine the accuracy and range of their specifications, this sensor has been verified. pH sensor is shown in Figure 9.10 and the overview of the process of pH sensor is shown in Figure 9.11.

9.6.4 Humidity and Temperature Sensor

Humidity is the amount of air vapor from water. Moisture sensors detect the relative humidity of the environment surrounding them. DHT11 evaluates humidity and atmospheric temperature. By calculating the electrical resistance between the two electrodes, the DHT11 calculates relative humidity. It expresses relative humidity as a percentage of the air humidity ratio to the maximum amount that can be held at the current temperature in the air. Moisture sensors use capacitive methods to find out how much moisture is present in the air. Measurement of temperature is done using the thermistor principle. This sensor provides stability and high reliability over the long term. It also offers the benefits of high-cost performance, quick results, and quality satisfaction. This sensor can also monitor moisture, as it highlights a precise calibration of the calibration chamber for humidity. Figure 9.12 shows DHT11-temperature and humidity sensor.

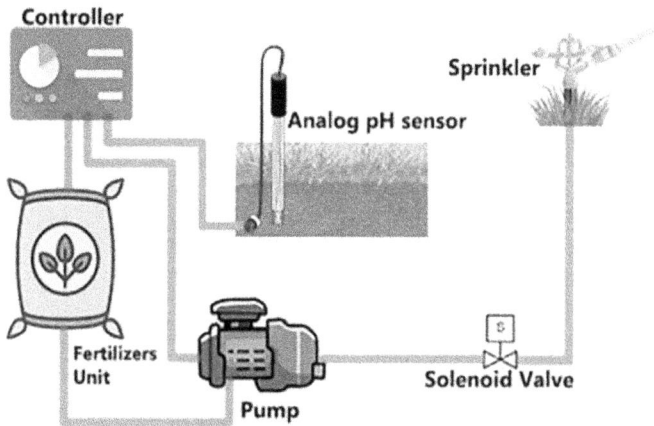

FIGURE 9.11 Overview of the process of pH sensor.

FIGURE 9.12 DHT11-temperature and humidity sensor.

9.6.5 MODULES

9.6.5.1 Arduino Mega 2560

The Arduino Super 2560 is designed to build robots based on Arduino and to do research based on 3D printing technology.

Technical specifications: ATmega2560 is the basis of the Arduino Mega 2560. It consists of 54 optical pins for input/output, 16 analog inputs, 4 UART pins (Universal Asynchronous Receiver and Transmitter). You should easily use the USB port to connect to the PC.

9.6.5.2 ESP 8266

The ESP8266 Wi-Fi Module would be an optimized TCP/IP protocol stack SOC that provides Wi-Fi network connectivity to every microcontroller. The ESP8266 platform supports APSD for VOIP applications and Bluetooth co-existence interfaces and is a cost-effective module.

> **Technical Data:** 802.11b/g/n; Wi-Fi Direct, 1MB Flash Drive, SDIO 1.1/2.0, SPI, UART, <1.0 mW Standby Power Consumption.

9.6.5.3 Breadboard BB400

The BreadBoard-400 is a solderless breadboard with 400 points of attachment, i.e. 400 points of insertion of cable. The BB400 has an IC circuit area of 300 tie-points plus four 25-tie-point power rails. The housing is made of white ABS plastic with rows and columns of written numbers and letters.

> **Functional Specifications:** 36 Volts, 2Amps, 400 points of tie, 50000 points of insertion.

9.6.5.4 Breadboard Power Supply

Control Module for the breadboard MB102.

> **Functional specifications:** 5v or 3.3v compatible, Output voltage: 5v and 3.3v, Max output current: <700 mA; Arduino, AVR, PIC, ARM, ARMA compatible.

9.7 PROPOSED SYSTEM

Current developments in electronic device miniaturization and wireless communication technology have led to rise of an energy-efficient WSN. This allows for the timelier, accurate, and convenient acquisition of field information. Figure 9.13 shows an overview of an IoT system; sensors are used for monitoring the environment to generate the data. These data can be processed locally or uploaded for secure storage to the cloud center. The devices could be interconnected to one another for communication and can interconnect with the internet. Microcontrollers, as a smart service system, make all the devices work together. Software and hardware are the components in microcontrollers. For processing, the sensing data will be sent to the microcontroller, and the actuators generate signals. In this way, the entire system is controlled toward the destination state. To end-users, the IoT system should offer intelligent services and should be appropriate interactions among them.

The PH sensor in this system is ideal because it has specifications that fit the proponents' standard for monitoring soil acidity. This sensor delivers reliable and accurate information. The pH sensor is used to sense the soil's nature. Details obtained

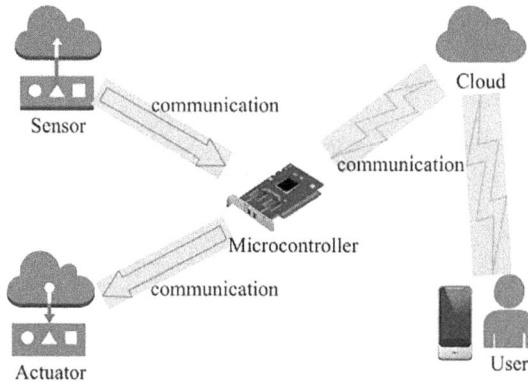

FIGURE 9.13   General IoT system.

are sent to the cloud and the data processed; details of the required fertilizer are sent to the mobile phone customers to produce a good yield. The LDR sensor detects the presence and absence of light, and for some crops, the intensity of light is an important factor for growth and production.

Nowadays water is scarce. Water wastage should be avoided while irrigating. So the plants or crops should only be irrigated when it needs to be. When plants transpire more water, relative atmospheric humidity increases. This presence of a large quantity of relative humidity increases the chance of attacking a disease. Soil moisture levels in the field require periodic inspection, from where one can determine when to do the next irrigation and how much water should be used.

For this, it is using soil moisture sensor and temperature humidity sensor, volumetric water content in soil, and the amount of moisture and heat present in the air are measured by using these sensors, respectively. These data are sent to cloud and will be compared with the predefined data, analyzed, and the resulting data will be sent to the mobile client and the commands will also be passed on to the motor. When the water level in the soil is less and the temperature is high, the motor will be turned ON and turned OFF automatically when the water level in the soil reaches a certain limit. Thus in this system, the whole process of irrigation is automated. Workflow of transmitter and receiver is shown in Figures 9.14 and 9.15, respectively.

9.7.1   SMART AGRICULTURE

In this new era, farming can be achieved using many of the new innovations. WSN is used here for low-cost and high-yield crop production. Humans are not interested in agriculture nowadays. Networks for human wireless sensors are used to decrease the initiative. Here, sensor nodes gather data and then send it to the farmers and specialists in agriculture. The use of some extra software and hardware information is conveyed to mobile phones. Farmer is able to control cell phones anywhere at any moment. Many farmers may be classified into this application and even the expert. This is more fit to the agriculture-dependent nations such as India [24].

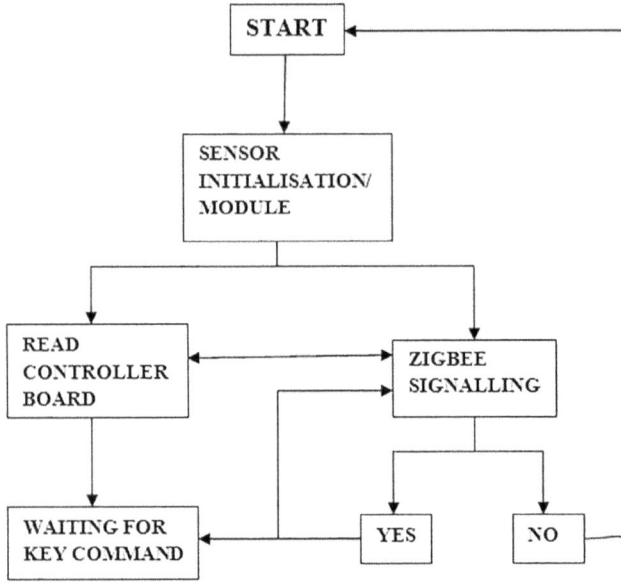

FIGURE 9.14 Workflow of transmitter.

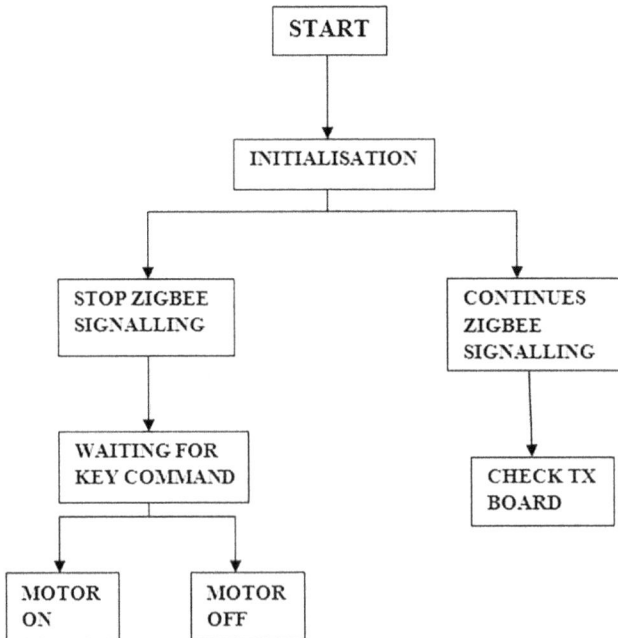

FIGURE 9.15 Workflow of receiver.

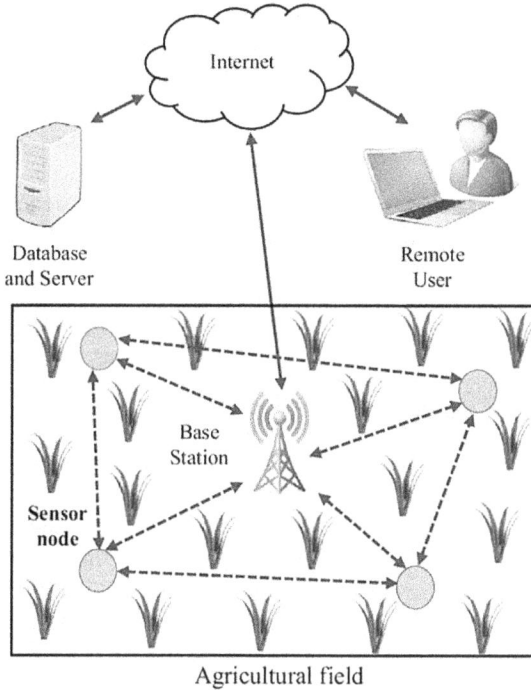

FIGURE 9.16 Smart agriculture fields.

By planting crops without infections and at precise soil water content, the smart farm enables the farmer to produce higher benefit. It decreases human activity due to the automated process and views crop growth from a mobile phone. The expense of installation is minimized by wireless communication. In order to communicate the data to farmers, internet access is needed at all times. The predefined forecast of the weather situation helps farmers to grow crop depending on the weather condition.

Instead of using conventional agriculture in this modern term, smart agriculture field in Figure 9.16 is achieved by the farmers. WSN are used to track crops in this article. Farmers can calculate the amount of water, moisture content, temperature, and even diseases in crops that are affected. The relevant data information is gathered by the sensors and processed on web servers. Then the processed information is transmitted automatically to two registered phone numbers.

9.7.2 DIFFICULTIES IN SMART FARMING

The fundamental measure of the dispersion of innovation in agriculture is that land ownership is too limited, hurting the growth of long-haul profitability. Similar to high yielding crops, all of our advancements are for watered fields, while 48% of our cultivated region is dry land. As per the 2016 Agricultural Census, 80% of land ownership is below 2 hectares and only 45% is the full edited area. Almost 90% of ranchers are remote and insignificant.

At present, the usual size of a homestead is merely 1.15 hectares. Just 5% of ranchers work more than 4 hectares onshore. The early beneficiaries were ranchers who had the opportunity to pool in their lands to build their homestead scale to 100–200 pieces of land in any case. Paradoxically, only 5% of ranchers operate on land above 4 hectares. Those misusing savvy inventions are mostly not ranchers, but massive agro-organizations. A portion of these instruments are used for hazard control by ranch advance organizations, the industry must beat growing water deficiencies, restricted connectivity of fields, and ripeness of terrains that are difficult to track costs.

Furthermore, current methods are inadequate to resolve the problems. In the world of tiny embedded objects, security problems must be anything but challenging and realistic to implement.

9.8   CONCLUSION

The proposed system offers several advantages, and it can operate with fewer people. Conventional water irrigation requires large quantity of water that leads to water scarcity. This creates uneven irrigation. The smart farming systems guarantee high productivity with water-efficient use. And then when the soil moisture is below the reference value does the device provide water. Due to direct flow of the water to roots, water saving happens and also helps to preserve a relatively stable soil moisture ratio at the root zone. Automation of irrigation control mechanism using appropriate environmental parameters that have been observed. The agricultural monitoring system based on IoT serves as a reliable and efficient system for monitoring environmental parameters effectively. The system is designed to help farmers improve. Uses of the agricultural monitoring system based on smart sensors have been used to increase crop yields by monitoring environmental constraints and giving customer information. Wireless field monitoring will enable users not only to reduce manual labor but also to see precise changes in it. That kind of system can be installed and maintained easily. In the proposed system, the scope for future work will include manufacturing, experimental investigation, and data analysis to predict crop yield, control solution, and complex network setups.

9.9   CHALLENGES AND FUTURE SCOPE

Agriculture has been addressing global problems such as lack of irrigation system, climate rise, groundwater density, food shortages, waste, and much more. To a great degree, the destiny of agriculture rests on the acceptance of different cognitive solutions. Although research on a wide scale is still under way and some implementations are previously present on market and industry remains extremely underserved. Farming is also at an early stage when it comes to dealing with practical matters, faced by the farmer and to counter them, use automatic decision-making and statistical solutions.

Applications need to be more stable in order to exploit immense scope of the artificial intelligence (AI) in the agriculture. Then only it will be able to manage regular adjustments in environmental circumstances, promote decision-making in real time and make the use of an adequate platform/framework to effectively capture qualitative data. The exorbitant expense of numerous cognitive strategies employed in farming industry is another significant factor. To ensure the technology hits the people, solutions need to become more accessible. The solution will be made more affordable by an open source network, resulting in faster acceptance and greater penetration among farmers. The technology would be useful to help high-yielding farmers and to provide a stronger seasonal crop in the regular intervals. Lots of countries like India, depend on the monsoon farmers for their farming. They rely primarily on predictions of the weather conditions from different departments, particularly for the rain-fed farming. The AI technologies can be beneficial in forecasting the weather and other agricultural conditions, such as groundwater, soil quality, pest attack, and crop cycle, etc. With the assistance of AI technology, the specific prediction or forecast would reduce much of the farmers' concerns.

AI-driven sensor is very useful for extracting valuable agricultural data. In optimizing efficiency, the data would be useful. There is massive area for these sensors in the agriculture. The data such as soil quality, groundwater level, weather, etc. may be obtained from agricultural scientists; these would be beneficial for improving the cultivation process. In order to get the details, AI-empowered sensor is also being mounted in robotic harvesting equipment. AI-based advisories are hypothesized to be beneficial in raising production by 30%. The greatest threat to farming is the disruption to crops due to some form of catastrophe, even pest attack. Farmers lose their crops much of the time because of the lack of proper knowledge.

The technology will be useful for farmers to defend their agriculture from some kind of the issues in this cyber era. In this direction, AI-enabled picture recognition would be beneficial. Many businesses have deployed drones to the monitor productivity and to detect some kind of the pest attack. These practices are fruitful several times, offering the incentive to have the mechanism to track and to protect the crops. A robotic lens zooms on a tomato seedling's yellow bud. Photos of the plant flow through algorithm of AI that calculates just how long does it take for flower to become mature tomato prepared for a grocery store's picking, packaging, and production portion. The technology is researched and developed at Nature Fresh Farms, a 20-year-old company that grows vegetables on 185 acres between Ohio and Ontario.

Keith Bradley, Nature Fresh Farms' IT Boss, said that knowing how many tomatoes would be ready for resale lets the sales staff's job uncomplicated and directly benefits the bottom line. It's only one example of the AI transforming the agriculture, an emerging pattern that is spurring the agricultural revolution further. AI will help mankind face one of the greatest problems, from identifying diseases to forecasting which crops can produce the highest returns: feeding extra two billions of people by 2050, as the climate change disrupts rising seasons, the arable land is turned into the deserts filling once-fertile deltas with the seawater. By the middle of century, the United Nations predicts that we will need to boost food demand by 50%. Between 1960 and 2015, food production tripled as population of the world increased from three billion people to seven billion. Although technology has played a part in fertilizers, chemicals, and computers, more benefit is traced to only plowing lots of land, diverting the fresh water to farms, cutting trees, rice paddies, and orchards. It will have to be more ingenious this time around. In the next few years, AI is likely to change the market and agriculture. Technology is useful for farmers to learn different varieties of the hybrid crops that will provide them with more revenue within a short time frame.

The correct introduction of the AI in the agriculture would help the process of cultivation and then create the environment for the market. There is a massive excess of food around the world, as per the data from leading organizations and using the correct algorithm, this issue is also being solved, which not only saves time and resources but also contributes to the sustainable growth. Funded by exploiting technology such as AI, there are greater opportunities for digital transformation in agriculture. But, because of the manufacturing process that takes place once or twice a year, it all relies on the tremendous data that is very difficult to obtain. Farmers, however, are dealing with changing situations by introducing AI to add digital transformation to agriculture. This is only one example of the AI changing agriculture, the emerging pattern that helps to stimulate a revolution in agriculture. This time around, it would have to be more resourceful.

For different IoT-based applications, multiple hybrid architectures may be planned as future works. It can be realized by approaches focused on software and hardware. In addition, it is possible to propose artificial learning-based approaches like reinforcement learning, neural networks, game theory, and fuzzy logic. On the other hand, on the basis of methods of finding the shortest way, the issue of power efficiency is studied from several perspectives, such as routing and synchronization between devices.

ABBREVIATIONS

IoT	Internet of Things
WSN	Wireless Sensor Network
LDR	Light-Dependent Resistor
DHT	Digital Humidity and Temperature
GDP	Gross Domestic Product
GPS	Global Positioning System
AVR	Alf and Vegard's RISC processor
SMS	Short Message Service
Wi-Fi	Wireless Fidelity
MatLab	Matrix Laboratory
SDI	Serial Digital Interface
A/D Convertor	Analog to Digital Convertor
RFID	Radio Frequency Identification
NFC	Near Field Communication
Cds	Cadmium Sulfide
pH	Potential of Hydrogen
TX	Transmitter
AI	Artificial Intelligence

REFERENCES

[1] FAO, "AQUASTAT: Water uses," 2016, http://www.fao.org/nr/water/aquastat/water_use (accessed on 5 January 2019).

[2] K.R. Sarode, and P.P. Chaudhari, "Zigbee based agricultural monitoring and controlling system," *International Journal of Engineering Science and Computing*, 8(1), 15907–15910, 2018.

[3] L. Atzori, A. Iera, and G. Morabito, "The internet of things: A survey," *Comput. Netw.*, 54, 2787–2805, 2010.

[4] R. Venkatesan, and A. Tamilvanan, "A sustainable agricultural system using IoT," in *International Conference on Communication and Signal Processing (ICCSP)*, 2017.

[5] J. Amalraj, S. Banumathi, and J. Jereena John, "IoT sensors and applications: A survey," *International Journal of Scientific & Technology Research*, 8(8), 998–1003, 2019.

[6] B. Boman, S. Smith, and B. Tullos, *Control and automation in citrus micro-irrigation systems*, Gainesville: University of Florida, 2006.

[7] N. Gondchawar, and R.V. Kawitkar, "IoT based smart agriculture," *International Journal of Advanced Research in Computer and Communication Engineering*, 5(6), 838–842, 2016.

[8] K. Anitha, "Automatic irrigation system," in *2nd International Conference on Innovative Trends in Science, Engineering, and Management*, ISBN:978-93-86171-10-8, 2016.

[9] I.F. Akyildiz, W. Su, Y. Sankarasubramaniam, and E. Cayirci, "Wireless sensor networks: A survey," *Computer Networks*, 38(4), 393–422, 2002.

[10] A.J. Garcia-Sanchez, F. Garcia-Sanchez, and J. Garcia-Haro, "Wireless sensor network deployment for integrating video-surveillance and data-monitoring in precision agriculture over distributed crops," *Computers and Electronics in Agriculture*, 75(2), 288–303, 2011.

[11] C. Kamienski, J.-P. Soininen, et al., "Smart water management platform: IoT-based precision irrigation for agriculture," *Sensors*, 19, 276, 2019, doi:10.3390/s19020276.

[12] G. Sushanth, and S. Sujatha, "IOT based smart agriculture system," in *2018 International Conference on Wireless Communications, Signal Processing and Networking (WiSPNET)*, Chennai, India, pp. 1–4, 2018, doi: 10.1109/WiSPNET.2018.8538702.

[13] L.C. Nogueira, M.D. Dukes, D.Z. Haman, J.M. Scholberg, and C. Cornejo, "Data acquisition and irrigation controller based on CR10X data logger and TDR sensor," in *Proceedings Soil and Crop Science Society of Florida*, pp. 38–46, 2003.

[14] Y. Kim, R. Evans, and W. Iversen, "Remote sensing and control of an irrigation system using a distributed wireless sensor network," *IEEE Transactions on Instrumentation and Measurement*, 57, 1379–1387, 2008.

[15] A. Fanti, M. Spanu, M. B. Lodi, and G. Mazzarella, "Non-linear analysis of soil microwave heating: Application to agricultural soils disinfection," *IEEE Journal on Multiscale and Multiphysics Computational Techniques*, 6, 1–11, 2017.

[16] W. Huang, Q. Guan, and J. Luo, New optimized spectral indices for identifying and monitoring winter wheat diseases, *IEEE Journal of Selected Topics in Applied Earth Observations and Remote Sensing*, 7(6), 2516–2524, 2014.

[17] S. Jagannathan, and R. Priyatharshini, "Smart farming system using sensors for agricultural task automation," in *Technological Innovation in ICT for Agriculture and Rural Development (TIAR)*, IEEE, 2015.

[18] P. Pawar, V. Turkar, and P. Patil, "Cucumber disease detection using artificial neural network," *IEEE Sensor Journal*, 10, 212–216, 2015.

[19] L. Gupta, K. Intwala, and K. Khetwani, "Smart irrigation system and plant disease detection," *International Research Journal of Engineering and Technology*, 4(3), 80–83, 2017.

[20] K. Shinghal, and S. Neelam, "Wireless sensor networks in agriculture: For potato farming," available at SSRN: https://ssrn.com/abstract=3041375, 2017.

[21] S. Lachure, A. Bhagat, and J. Lachure, "Review on precision agriculture using wireless sensor network," *International Journal of Applied Engineering Research*, 10(20), 16560–16565, 2015.

[22] D.N. Suma, S.R. Samson, S. Saranya, G. Shanmugapriya, and R. Subhashri, "IOT based smart agriculture monitoring system," *International Journal on Recent and Innovation Trends in Computing and Communication*, 5(2), 177–181, 2017.

[23] C.N. Verdouw, S. Wolfert, and B. Tekinerdogan, "Internet of things in agriculture. CAB reviews: Perspectives in agriculture, veterinary science," *Nutrition and Natural Resources*, 11(35), 1–12, 2016.

[24] D.D.K. Rathinam, D. Surendran, A. Shilpa, A.C. Grace, and J. Sherin, "Modern agriculture using wireless sensor network (WSN)," in *5th International Conference on Advanced Computing & Communication Systems (ICACCS)*, Coimbatore, India, pp. 515–519, 2015, doi: 10.1109/ICACCS.2019.8728284.

10 Applications in Automobile Industries

Warehouse, Logistics and Delivery Systems, Mobile Robots

G. Sathish Kumar

Sri Krishna College of Engineering and Technology, Coimbatore, India

D. Prabha Devi, R. Ramya, and P. Rajesh Kanna

Bannari Amman Institute of Technology, Erode, India

CONTENTS

DOI: 10.1201/9781003181668-10

10.1   INTRODUCTION

Automobile manufacturing is one of the chief economically significant and technologically complex industries in the world. In this modern era, all the automobile manufactures need to implement novel strategies to increase their product range and propose highly personalised products to be in the competition. This strategy builds an application and forces to transform the industries to be more flexible and agile in the production process. The new industrial revolution has taken place in the past years due to the automation technologies like artificial intelligence (AI) or machine learning in the automobile industries. The collaborative work of robots, robot arm and the Internet of Things (IoT) coupled with the AI has produced huge portion of automobile chassis, power trains and many other components. It greatly minimises the work and efforts of the human workers. In many automobile industries, the most complex tasks are allowed to be handled by the AI-enabled robots. These robots are completing the task much faster than the human efforts. Robots outperform the works with greater accuracy and precision and result in increasing productivity.

Most of the companies use the augmented reality and virtual reality to focus on the manufacturing issues. These technologies have minimised the cost of error during the production stage. The automated automobile industries have become the benchmark for other industries. They are visualising and transforming the plants into a high-class manufacturing hub along with the digitalisation. The digitalisation in the automobile industries results in highly transparent, highly visual and much organised way in industry development. The business needs in the automotive industry is to create the optimum way for the faster product development and quick adaptability. The use of autonomous mobile robots (AMRs) is the clear tendency in the automobile industries. The AMRs do not require the magnetic strips for the guidance for the mobility inside the manufacturing unit. Instead, they require only the internal maps for the navigation, and it is very easy to update. The autonomous robots are mainly used in robotic vision, spot and arc welding, assembly, painting, sealing, coating, machine tending and part transfer, materials removal and internal logistics.

The advanced hardware architecture and control software components permit the autonomous operation even in the dynamic environment. In the automated guided vehicles (AGVs), the central processing unit takes the entire control for performing the scheduling, routing and dispatching of the decisions. In addition, AMRs can interconnect and communicate with add-on resources like machines and systems to distribute the decision-making process. The distributed decision-making process creates the system to act dynamically for all kinds of state and environment. The AGV creates the trend in automation of material shipping. Initially, these AGVs need static routes with marginal on-board intelligence. This drawback limited the usage of AGVs in the area where the pick-up and drop-off point of the material varied. The AMRs are replaced for AGVs. The AMR has the on-board intelligence with real-time adaptive skills. This AMR supports for the improvement market demand in flexibility and agility in connection with the variation towards the process and products.

The logistic process in the industry requires the synergistic optimisation methods for all the physical processes. The innovative technological applications in the automobile industries are in the need of drones or unmanned aerial vehicles (UAVs) to achieve the complex and efficient production. The UAVs can be controlled autonomously by the system with the pre-programmed operations or by the pilot on the ground using the remote control. Micro aerial vehicles (MAVs) with multi motors are used for autonomous inventory-taking in the huge warehouse. This MAV operation in the complex warehouse requires the obstacle detection, real-time assessment, image processing and mapping with navigation scheduling.

With the automation techniques and resolutions, the employments on the floor in automobile industries are safer. The automation yields higher production rates and higher productivity with superior quality in product development. The improved process control makes more effective use of materials, resulting in very fewer scrap.

10.2 LITERATURE REVIEW

The study on autonomous vehicles like driverless cars has become more renowned in the preceding 15 years. It gained more attention by the industries than any other technologies. The pedestrian recognition and tracking system has become one of the foremost concerns in the autonomous vehicles. In the autonomous collision avoidance system, the autonomous vehicles work very close to the pedestrians, tracking and recognition of ramblers. These are the fundamental operations in the self-driving vehicles. In addition, the self-driving vehicle should timely predict the immediate upcoming positions of the pedestrians. This feature helps to estimate the real-time route scheduling of self-governing vehicles [1].

Object detection and recognition is the major challenge in autonomous vehicles. The detection and localisation of the static and dynamic images creates the problem due to the higher variance in the object appearance. These variations are caused not only from the viewpoint and illumination but also due to non-rigid distortions, and intra class patchiness occurs in the shape and supplemented graphical properties. These distortions can be made more elegant by using the rigid templates and the bag-of-features for object detection [2].

The current logistic systems have become the significant means to increase the efficacy of the material flow and reducing the distribution capital in many industries. In the due time, the present development of the E-commerce industry also contributed to the growth of demand for logistics and stimulated the technology development for logistics in the market [3]. The automation technology has more applications in logistics. Some of the applications using automation technology are done using the warehouse control system (WCS), industrial robots and programmed storage systems. The WCS is applied to govern the warehouse equipment. WCS are constructed using software interface model for efficient governing [4]. The industrial robot performs its function based on multiple axes with automatic programmable position controlled architecture. It handles all the resources such as tools and parts using the variable programmable procedures to accomplish the range of tasks. The industrial robots can be made and programmed to fit to the need of automobile industries. In real time, the industrial robots are mainly used in trailer and container loading, stationery and mobile tools handling, home delivery and customisation [5].

The work-measurement technique has been carried out in the dispatch section of the industry. The process flow standard should be followed to overcome the unnecessary movements and delay by the automated vehicles during the dispatch. It will greatly reduce the complete fabrication cost and regulate the whole deviations in the product and process [6]. The autonomous robots are greatly employed in assembling the parts of the unusual weight. Because of their usage, these robots are exhaustively used in all the autonomous assembling units. The less life cycle and diverse client requirements have triggered the autonomous robot developers to advance the ability and efficacy during its manufacturing exercise. The frameworks used in the process of assembling will certainly produce the items with less expenditure and greatest calibre. It produces the items as conceivable amount of time to meet the client's schedule. Important feature of the framework is that, it has a preference to amend or respond quickly to the variations in element structure and element appeal without any substantial speculation [7].

By detecting the important features of automobile component sectors, the supply chain management has turned to a key concern. It has resulted in the identification of new autonomous supply chain to meet the needs of self-motivated business environment [8]. The remanufacturing mechanism and the product recovery mechanisms gained its popularity during the recent years among the automobile industries owing to the commercial, environment and legal ethics. Circular economy model has been introduced due to the best utilisation of the resources with recent manufacturing methods and limited landfill space. The role of circular economy in the factory is to combine the process of manufacturing and remanufacturing [9].

Numerous companies have constructed the world class manufacturing (WCM) system to compete with the high-class industries and to improve the production performance. The huge amounts of data and information have been used by the automotive industries to produce the flexible and agile systems for decision making. The continuous improvement and cost reduction are the key achievements of WCM. The automotive industries produce and offer many personalised products in the markets and move towards the mass production [10]. The smart factory systems with cyber-physical production system (CPPS) are implemented by many automobile industries to fulfil the explicit production features and the scheduling constraints. On the contrary, the cloud-based information system is employed in many medium-scale industries for the smart manufacturing services. The advanced planning and scheduling systems behave like a batch for enterprise resource planning systems to solve the shop floor level problems [11].

The concepts of smart manufacturing, industrial internet and integrated industry have the great possibility to disturb the complete industry by renovating the way goods are designed, fabricated, distributed and paid. The integration of cyber-physical systems and IoT into the logistics has empowered the tracing of material flow with upgraded transportation and risk management [12]. The reverse logistics approach has acknowledged the considerable vision during the past few years due to the mixture of economic, environmental and social factors. The first level of reverse logistics is to collect products and it ends up with re-processing of the product during the manufacturing level. In the reverse logistics, the higher supplier chain results in uncertainty in the product [13].

The AGV has been used widely in intra-logistics and material handling operation in the automobile industries. Along with the AGV, the AMRs are constantly combined and used in the industrial areas to perform and complete the given tasks. These robots are in the form of robotic arms and actuators. This setup has been built on the top of mobile podiums. The main use of AGV and AMR is to transport the resources around the manufacturing floor in the warehouse. To gain the superior performance on the factory floor, the AGV and AMR fleets need to be optimised. Robots toil together with the factory workforce in the assembling unit to assemble the products. Theses robots are used around the world. The advanced AGV and AMR systems can be employed robustly and unified with complete industry production for accomplishing the smart manufacturing [14].

10.3   INDUSTRIAL APPLICATION OF MOBILE ROBOTICS

Over the years, robotics development concentrated mainly on the parameters like performance, accuracy of repetition along with its speed. Some additional criteria which have greater importance must also be focused. This criterion aids in the process of industrial transformation. In the field of manufacturing, the basic expectation from the modern robots is that, they must be flexible, adaptable and independent. These criteria evidently prove that the development and the progress in industry depend on the automation technology and mobile robotics. It is not the case that the industrial field alone is benefiting from this innovation. Certainly, it will have some impact in our day-to-day life in many other ways in the near future.

Classic robots function absolutely on the basis of a programmed pattern. They do not have independency. Further, they do not require any sensors too. Mobile robots emerged out of this existing system with enormous differences. The major difference can be explained by associating a traditional robot with an industrial robot. Depending on the application, the mobile robots are employed with a set of sensors which plays a vital role in the process of transformation. The sensors may be a camera, ultrasound detector, laser scanner, etc. Upon integrating these components, mobile robotic system could be able to respond instinctively for all the events.

Another basic difference may be the method of programming. A mobile robot is designed depending on the roles and the tasks allotted to it. It runs with a software program which comprises sequence of commands. With the presence of sensory system and autonomous programming, a mobile device can perform independently. Normal robots repeat the same process whereas a mobile robot changes its movement immediately once it encounters any changes in the environment.

The conventional industrial robots operate within a small range, and hence they are immovable. Mobile robots have an independency to find their own varying novel environments based on their own perceptive. They can act as an autonomous device as they operate with its own learning ability in agreement with the surroundings. Once any novel information is identified by the mobile robot, it is exposed to further processing. This is considered for future sequence of the same motion in the appropriate surrounding. Mobile robots may respond immediately to all the events and in some situations they find alternative solutions for unusual errors too. Normal robots simply throw an error message if they are under the same challenging situations.

Mobile industrial robots are designed in such a way that they can handle any sort of industrial setup. By nature, these robots are used mostly in workbench applications, but now, mobile industrial robots lead to a novel method of manufacturing. They are allowed to perform mobile tasks like product delivery because of their advanced controls and robotics. With this, a company saves the time of manufacturing process thus ends up in a less cost final product.

Even though mobile robotic technology has the power to rule a large number of varying industrial sectors, it may have few cons with it. Manufacturing sector may be modernised by enabling robots to move independently in a variety of work areas. Robots may work along with the humans which in turn reduces the demand of labour. Mobile robots have the capability to assist even in the medicinal and the surgical field. But there are few drawbacks in this. From the perspective of movement of robots based on their position, it is difficult to have a perfect coordination as suggested by Zhang et al. [3]. Human safety must be taken into consideration as a robot may malfunction during a manufacturing process or anywhere inside the environment. The malfunctioning may end up in less production. So the robotic design should be carried out in such a way that it must prioritise the safety of human resources apart from its programming. Further, they must be able to coordinate with multiple independent robots. Specifically in the field of surgery, there is no space for any error from the robot's side. Regardless of these drawbacks, mobile robotic technology modernises various aspects across different industrial sectors as recommended by George [15].

10.3.1 HISTORY

D.S. Harder, an Engineering manager at Ford motor company initiated the term automation around 1946. Initially, automation means the increased availability of some automatic devices in manufacturing and production. But now, automation is used to replace human interference in most of the industries. Gradually, improvements in this area have become highly dependent on the advanced technologies and the development of processing power as mentioned by Groover [16].

At present, industrial robots are driven with mechanical arms which make it to perform humanlike actions. Improved technologies in the areas of sensors, mathematical control and computer miniatures lead to a major impact on the feedback control system which acts as a key for robotics as provided by Groover [16]. In 1962, the General Motors factory located at New Jersey, USA, first used industrial robot for spot welding and die castings. Later on, the field of industrial robotics was explored within a large scale of manufacturing industries. A lot of new companies came into the picture. In 1973, a company named Kuka evolved, followed by a number of companies like Nachi, Yaskawa, ASEA and so on. By the end of 1980, it was found that a new robotics company emerged every month.

Nowadays mobile robotics has expanded a lot due to its importance and reliability in most of the industrial setup. The frequency of the mistake committed by a mobile robot is negligible when compared to the mistakes committed by the human resources. Mobile industrial robots are very easy to operate. Because of its simplicity, it plays a vital role in a majority of industries. Mobile industrial robots make use of advanced technology, yet it is well understood by the human resources who can work along with the mobile robots. With these advantages, the mobile robots can operate endlessly and it will never come up with any complaints about the working hours. This in turn increases the efficiency in any manufacturing industry. The only disadvantage with this mobile robotics as of now is its high cost of recovery from failure. This in turn results in the delay of the end product. That is the reason why most of the industries are not providing major responsibility to mobile robotics. They simply try to do all smaller tasks with mobile robotics with which the industrial sectors may reduce the man power to a certain level as suggested by Zhang et al. [3].

In flexible robotics, the next level is to increase the mobility. In manufacturing, fixed robots have a same place and they perform a single task. With added flexibility in application, mobile robots can operate on varying applications. The applications of mobile robotics include security, medicinal and surgical purpose, and warehousing, personal assistance, space and ocean exploration and so on. We list some of the industrial applications of mobile robotics in Section 10.3.2.

10.3.2 INDUSTRIAL APPLICATIONS

10.3.2.1 Arc Welding

Arc welding is otherwise known as robot welding. It is the process in which the mobile robots are involved in fully automating the process of welding. The mobile robots can handle both the weld and the parts involved in that process. Arc welding is a process where two metals are joined using electricity. The purpose of electricity is to create heat to melt the metals. After heating, metals are cooled down which results in a final welded metal.

Robotic welding is used for resistance spot welding along with arc welding. Spot welding is a type of electric resistance welding where two metal sheet products are joined by the heat which is obtained through the resistance to the electric current. These mobile robots are used in automotive industry which involves a high level of production applications.

The concept of robotics was introduced in the year 1960s. But the usage of mobile robots in the field of welding started in the automotive industry in the late 1980s. From then, the number of robots and the number of applications involving the mobile robots increased drastically. As the human safety is much bothered, around half of the mobile robots existing in the year 2005 are meant for the purpose of welding as mentioned by Cary and Helzer [17].

Arc robot welding grows rapidly as it covers around 20% of the applications of the industrial robots. Manipulator and the controller are the major components of such robots involved in the process of arc welding. The controller is called the brain of the robot. The movement of the robot is accompanied by the manipulator. Based on the design, different systems direct the arms of the robot in different directions. Selective Compliance Articulated Robot Arm and Cartesian coordinate robot are some common types.

In arc robot welding, the robot either welds an already programmed position or it can weld a position based on the guidance of automatic inspection. In some cases, both may be collectively used as suggested by Turek [18]. With the advancement of robotic welding, the equipment manufacturers improve their accuracy. In late 1990s, signature image processing evolved in order to optimise the concept of welding.

10.3.2.2 Materials Handling

Robotic material handling systems are used in a variety of industrial sectors. Material handling is used to transfer the objects with the robot's ability. Robotic arms move the parts over the conveyor belt or sometimes it is used to hold a part of production and place it in some other place. Robots must be fixed with a perfect gripper with which they can accurately and efficiently change the position of the product from one to another. Such material handling robots are in very high demand because of their wide range of applications.

The benefit of robotic material handling lies with its uptime. Manual process of handling materials consumes much of the time. It is unreliable and unproductive. But robots can continue to work without any breaks and they provide good consistency. These robots need a minimal downtime for the purpose of maintenance. Along with the high uptime, the robotic material handling completes the task in a much faster manner. This in turn creates a positive impact on the production. The benefit of having huge uptime and production compounds over time helps manufacturers to a greater extent.

Robots employed in material handling are mainly used to move the product from one place to another, to pack the end products or to pick up the flawless product. Further, they can automate the process of moving a part of one machine into equipment. Labour cost is reduced and the human safety is also considered as the risky activities are performed by the robots.

10.3.2.3 Machine Tending

Machine tending means loading or unloading a particular machine with its sub parts or materials. Human resources are meant for performing the task of machine tending. Few machines in the industrial sector should be tended by human employees who keep the raw parts into the main machine. Once the machine completes its task, the worker needs to remove the raw parts from the machine. Qualified employees are required by the industries in order to perform this process of machine tending. But due to the lack of such well-qualified employees, industries are moving to robotic machine tending to replace the shortage of workers.

A robotic machine tending process can be done repeatedly if the robot continuously receives raw materials. In the industries where the production is being done continuously, these robots play a vital role in reducing the cycle time and it helps in running the process without any break.

The advantages of machine tending includes increased flexibility when compared to static automation, maximum throughput with greater speed and performance, higher system uptime, servicing capability for multiple machines, minimum cost of operation, improved quality, capability to do all secondary services and greater worker gratification.

In most of the production industries, space is a major concern. So, many of the machine tending robots occupy less space. The major concern here is, where to place these machine tending robots? Flexible mounting is one of the answers. In flexible mounting, the arrangement of machine tending robots with the machine can be customised. One of the valuable ideas is to abide the machine tending robot inside a machine tool itself.

Machine tending robots have the capability to change the tools in a faster manner. There are a lot of tool changing and mounting options available. Some specific machine tending robots are well trained to act as a tool changer. These robots have the ability to work with a wide variety of tools irrespective of their weight. Robotic machine tool loading is very easy to implement and to proceed further. It extends its service in the area of palletising. Palletising is the process in which the products or goods are placed on a pallet for shipment.

Machine tending robots are highly useful in the area of injection moulding. Injection moulding is a process involved in manufacturing to produce parts in the method of injecting molten material into a mould. This process of injection moulding is done by the robots without any damage to the machine or to its spare parts. Robots perform the functionality of removing the part of a machine, adding inserts, cutting, labelling and laser marking.

10.3.2.4 Painting

In automotive paint applications, robots have been used over the decades. Initially, hydraulic version of the robots was available to perform the process of painting. In the hydraulic version of robotic painting, the quality and safety is not ensured. The electronic version of painting robots comes into the market to overcome the drawbacks of hydraulic painting robots. These novel robots perform the task with more accuracy and the results are delivered in precise thickness.

During the earlier stages of development, the industrial paint robots were very costly and huge in size. But now, the prices have got reduced in such a way that a medium-scale company can afford the same painting robots as the one used by any large-scale manufacturers. The selection of electronic paint robots varies depending on the size. The configuration may vary depending on the environment. Around five to six axis motions are possible with painting robots. Three axes are meant for basic motion and a maximum of three are allotted for application orientation. A standard method is designed to set the distance and path for the sprayer available in the paint robots. This removes the chance of having error when the same process is done manually by a human. In order to improve the effectiveness and reliability of these industrial paint robots, some programmed painting apparatuses are paired with it. This automatic painting equipment adds some of the extra advantage like reducing the paint wastage and energy cost.

Industrial paint robots are mostly used by vehicle manufacturing companies where the detailed painting works on their cars are done in an efficient and reliable method. In order to apply paint on all the parts of a car, these industrial paint robots are designed with an arm which is having the capability to move in both horizontal and vertical fashion. In the area of aerospace and defence, finishing the final process with some consistent coatings plays a major role. Industrial paint robots are meant for this perfect finishing process to avoid corrosion. Further, it offers anti-static indulgence and acts as radar evasion stealth.

The commercial building industry uses aluminium extrusions in order to protect the buildings. These extrusions can be present in the panels of the buildings, doors and windows. The companies may end up in the issue of slim margins. Those companies are in need to strengthen the quality with a minimal cost. In addition to this, they should have a faster production and customisation. With the aim of satisfying most of the customers, the manufacturers of aluminium extrusions are making use of these paint robots to apply several coatings to enhance the protection.

Industries involved in manufacturing agricultural and construction-related machines make use of these paint robots. Such machines face substantial operation in their environment. Prevention from rust and extension of machine's life cycle can be done through multiple numbers of coatings. This task is not that much easy to do manually. An industrial paint robot provides the durable coating for this robust equipment involved in agriculture and construction. In addition to the above-said areas, the industrial paint robots are further used in the manufacturing of cookware, cosmetics, defence, aluminium panels, etc.

10.3.2.5 Picking and Packing

A pick and pack robotic system is used to pick up single or multiple objects at a time. Then it places the picked item inside a container. All these processes are completed in a much faster way when compared to a human resource. This in turn increases the production speed. In addition to the speed, these robots are used to place the exact object in the respective container. Robots are precise in pick and pack. It is most accurate as at times human worker may not. Human workers may have injuries when they repeatedly perform pick and pack. Robots do not have such issue which increases the uptime. The manufacturers ultimately end up in an increased production rate.

10.3.2.6 Assembly

A wide range of industrial assembly robots exist in the market to perform faster and perfect assembly. Assembly robots put various parts of a machine together with a greater precision rate. Three configurations of assembly robots are available: six-axis arm configuration, four-axis arm configuration and modern delta configuration. These robots are provided with a greater system of vision and sense. The assembly robots are employed in automotive area which includes the components like motors and pumps, computer accessories, electronic field, medical equipment and household appliances.

10.3.2.7 Mechanical Cutting and Grinding

Industrial sectors face a lot of difficulties in automating the process of machine cutting and grinding. With the introduction of the industrial robots, it is very easy to perform the functionality of machine cutting and grinding. If a human worker takes 90–120 minutes to perform a simple machine cutting, the same operation is done by a robot within few minutes.

10.3.2.8 Gluing and Sealing

Robots for gluing and sealing are offered in a wide variety of sizes depending on their capability to bear with the load. These automated gluing and sealing processes reduce the cycle time, and it cuts down the cost per unit. Error rate is reduced to the maximum extent, and it is easy to find the repairs. The robots ensure the perfect quality and efficiency in gluing and sealing when compared to the human workers. Further, they reduce the wastage and increase the flexibility by the robots' easy programming.

10.4 AUTONOMOUS MOBILE ROBOTS

Before introducing the topic of AMRs, let us understand the term "Autonomy" at first. Autonomy defines doing a specific action by understanding the situation or the environment and making its own decision without the intervention of human input. So why is it important to have Autonomy in Robots? Autonomous robots do not reside on human decisions, instead they observe the environment or the platform in which they are working currently and based on the programming done on them, they choose the right conclusion to be performed. So is it not interesting to know that autonomous robots are doing a great job.

10.4.1 So What Are Autonomous Robots?

Autonomous robots are the most intelligent machines that perform the correct task by observing the environment, take the right decision and then do its manipulation within that environment. It means human do walking, talking, and stopping the door and even many more activities such that the autonomous robots do the same by observing its own environment.

10.4.2 Control Scheme of Autonomous Mobile Robots (AMRs)

Over the past few years, AGVs are being used. These vehicles are widely used in the industry where it reduces the human efforts. In industries, if these vehicles are provided with the location upon where the products are to be placed with the help of human intervention, they do its job by which it works along with fixed routes only, whereas AMRs are used in a different way, where in the development of today's robots, focus is on real-time response. The ability to adjust the motions of robots to their environment by a true two-way partnership helps the robot for its mobility. Figure 10.1 shows the control scheme structure of the mobile robots.

10.4.3 Comparison between AGVs and AMRs

There are lots of differences between AGVs and AMRs with respect to their identifications of direction, tractability, ease of access and production.

10.4.3.1 Direction Identification

Both AGVs and AMRs navigate and operate in a different way. The way it gets operated differs wherein AGVs are more expensive when the directions need to be changed. Whereas AMRs are not much expensive and tractable because they do not require lots of changes in their working environment and are not more expensive in order to progress through new operations.

10.4.3.2 Tractability

- AGVs are used to operate the same kind of task whereas AMRs are used to handle multiple tasks.
- AGV weigh around 150 kg whereas AMRs weigh about 50 kg approximately.

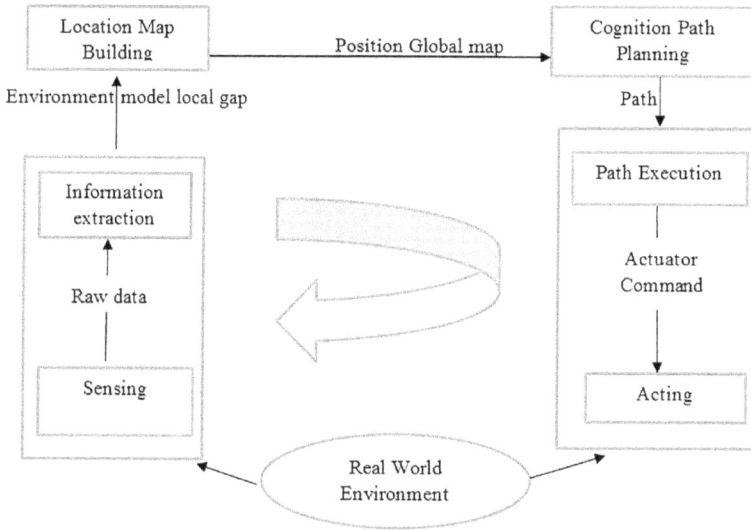

FIGURE 10.1   Control scheme of mobile robots.

- AGVs cannot identify the obstacles and need much care to operate it whereas AMRs can identify the obstacles easily if there is any, and they can easily navigate from the obstacles. So AMRs do not cause much damage and operate perfectly.

10.4.3.3   Ease of Access

To use AGVs, workers in the production company need prior knowledge in installing and training before use, whereas if AMRs are used in the production company, workers need no prior knowledge in operating it.

10.4.4   TYPES OF AUTONOMOUS MOBILE ROBOTS

There are four different types of driverless vehicles which can work with more efficiency and without human intervention, thus reducing the cost of human power in work environment.

10.4.4.1   Automated Goods Picking Machine

Goods picking machine really helps in reducing time to pick goods from one place and place the same onto the other area. They work in a way such that the machine is programmed with the location from where to pick and place the goods along with the sensors attached to identify the exact location as well as the obstacles. Also it reduces the work force of human power by avoiding the situations of unnecessary walking around the given environment. So it is easy for the warehouse manager to change whatever they want to do and whenever they want to do it.

10.4.4.2 Automated Driving Forklifts

Self-navigation is made in such a way that sensors to identify the products in the warehouse and the people's movement as well could be easily recognised. Navigation laser, 3D camera and sensors enable the safety indication of movements of these robots.

10.4.4.3 Automated Inventory Robots

An autonomous inventory robot provides a large set of prospects for observing and monitoring. When the products are to be placed in a specific place, to make things better by providing the RFID tag along with the product, these autonomous machines may automatically sweep out the product at specified places. Thus, these robots make the real-time data much easier for every two hours since they provide the data conditions for every moment in the warehouse.

10.4.4.4 Unmanned Autonomous Vehicles

In unmanned autonomous vehicles, sensors and controlling mechanisms empower crash expectancy and a natural plan that permits it to take on flight examples to interesting formats and to explore jumbled conditions.

10.4.5 PATH PLANNING IN AUTONOMOUS MOBILE ROBOTS

Most important property of AMRs is built with proper path planning to ensure that the product reaches to the right destination. Sensors are also fixed in it to identify the obstacles and if obstacles are found in the given path planned, then it starts to identify the optimal path to reach in the shortest time in the given environment. Two path planning approaches include local path planning and global path planning [19]. In global path planning, it is important for the robots to obtain the prior knowledge about the working environment to easily identify the alternate path if any obstacles are found. The process of deciding the optimal approach to take the robot from a starting point to a goal location is referred to as global path planning. Normally, when a robot is planning a local path, it is guided by a single straight line from the beginning point to the goal point, which is the shortest path, and the robot follows the line until it encounters an obstacle. The robot then executes obstacle avoidance by deviating from the line while also updating certain key information such as the updated distance between the present position and the goal point, the obstacle leaving point, and so on. To ensure that the robot can reach the destination precisely, this sort of path planning requires the robot to always know the position of the target point from its current position.

Pointbug algorithm [20] is used for better path planning. Pointbug is a newly created robot navigation system that navigates a robot point in a planar unknown environment loaded with fixed objects of any type. It calculates where the next point to go from a starting position to reach the target. The output of the range sensor, which detects a sudden change in distance from the sensor to the nearest obstacle, determines the next point. Inconstant distance readings are defined as rapid changes in range sensor output, whether increasing or decreasing. Since information about the surroundings can be collected instantly from the range sensor while the robot is moving, the algorithm can also work in a dynamic environment. The algorithm's performance is determined by the total number of sudden points found. As a result, whether it beats other bug algorithms is dependent on the environment's barriers; for example, if the obstruction is a circle, it will yield fewer abrupt points because a circle has no vertex. The total sudden points are affected by the total vertex in the obstruction.

10.4.6 APPLICATIONS OF AMRs

AMRs are widely used in many applications as follows:

10.4.6.1 AMRs in Distribution Centres

Mobile robots are most commonly used in warehouses and distribution centres. AMRs are widely used for palletising, loading, unloading, arranging and retrieving. Warehouse AMRs are more flexible and are used to carry out the individual packages even more than the bulk units.

10.4.6.2 AMRs as Disinfectants

As a result of the Covid-19 pandemic, the need for cleanliness and sanitation is now more important than ever in society. Customers nowadays want to know that the premises are sanitary. It is also beneficial for businesses to ensure that their buildings are clean and present a positive image, in addition to utilising disinfectants to combat disease germs. For effective sanitisation, an automated floor scrubber, as shown in the Figure 10.2, removes visible filth and can apply detergents and disinfectants. Some are equipped with sprayers to sanitise surfaces above the floor level.

FIGURE 10.2   Robotic vacuum cleaner.

10.4.6.3 Autonomous Security Robots (ASR)

ASR is widely used when it is important for the administrator to look over the industries without their presence. AMRs move here and there randomly in the specified environment and collect the data which can be fetched easily by the officers of the industry with the help of cameras fitted in that. These ASRs run in a dynamic way, but this can be modified and set up to remote devices such that they can view the suspicious activities at the specified places rather than the entire environment. It is also possible to sense the obstacles or the change in the environment with the help of autonomous navigation sensors in it.

10.4.6.4 Autonomous Robots in Healthcare

It is known clearly that autonomous robots are widely used in hospitals and healthcare centres for cleaning and to provide disinfectants. Since its usage is more efficient with replacement to the human, the spread of diseases dealing with virus and bacteria can be avoided and it is much safer to use in hospitals.

10.4.6.5 Autonomous Mobile Robots in Grocery Stores

In grocery stores, these robots are safe and reliable and move all over the stores and they have sensors to detect the stock materials. If the product is out of stock, then the message is collected by the grocery staffs and gets immediately filled in the required places. It also checks for other details like price incorrect details and other relevant information to the stock products. AMRs are also converted into use as shopping carts which makes the customers to easily get the products and scan each of the products that are collected.

10.4.6.6 AMRs for Hospitality

Hospitality business involved with conference places, hotels and restaurants uses AMRs as shown in the Figure 10.3. They not only provide the good hospitality by delivering the requested items but also decrease the labour intake, thus providing the higher profitability with the ease of these AMRs.

10.5 USAGE OF DRONES IN WAREHOUSE-BASED APPLICATIONS

10.5.1 Drones

In the logistics business, drones have shown a lot of potential. By 2027, this business is anticipated to grow by $29 billion with a compound annual growth rate of nearly 20%. Drones have shown to be quite useful in warehouses. Thanks to recent technological developments such as vision-based navigation and sensors, drones may now be utilised indoors [21]. Aside from improvements in drone technology, warehouses are expanding in size as global ecommerce grows. For example, drones can also be used for automated intra logistics and inventory inspections [22]. Warehouse managers may utilise drones to automate time-consuming and sometimes risky activities.

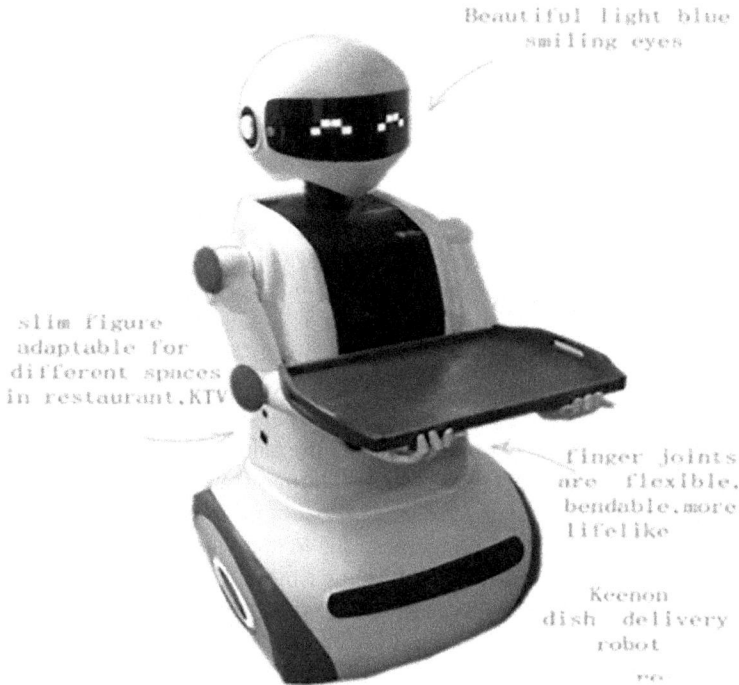

Beautiful light blue
smiling eyes

slim figure
adaptable for
different spaces
in restaurant,KTV

finger joints
are flexible,
bendable,more
lifelike

Keenon
dish delivery
robot

FIGURE 10.3   Robots designed for restaurants.

10.5.2   DRONES IN WAREHOUSE-BASED APPLICATIONS

Drones have become more frequent at warehouses in recent years. Large warehouses are investing more in automation and robots to boost efficiency. Warehouse operations account for 30% of overall logistics expenditures [23], so this isn't out of the ordinary. Furthermore, the difficulties in rising customer service demands, finding skilled employees and the development of e-commerce have all increased the need for warehouse operations to be more efficient.

The fourth industrial revolution has an impact on warehouses as well. They grow more computerised and networked, as in "warehouse 4.0". The modern scanning technologies like QR codes, AI, bar codes and RFID technology makes the drone-driven warehouse automation is now conceivable. Due to on-board computing power and efficient algorithms, scalable drone applications may also be created [24].

Warehouses, on the other hand, have a diverse and complicated structure that makes a drone program difficult to implement. Aspects to consider include layout (e.g. shelf, pallets, and boxes), geographic location, and kind of stored products, technology and size. Warehouses are used for a wide range of purposes. In terms of how raw materials and finished items are handled, distribution warehouses differ from cross-docking and production warehouses. Drones have become an increasingly important element of warehouse automation in recent years. Drones are intriguing because they can independently fly and hover, navigate indoors, land accurately, and even operate in fleets. Item intra-logistics, inventory management, surveillance and inspection are the three most potential indoor drone use cases in warehouses.

10.5.3 INVENTORY MANAGEMENT

In the realm of drones, inventory management may be used for inventory audits, cycle counting, inventory management, buffer stock maintenance, item search and stock taking. The practice of physically counting the number of things held in warehouses is known as stock taking. A stock take is usually done once a year or at the conclusion of the fiscal year. A way of counting a section of a warehouse's inventory more often is cycle counting. A small trained team of inventory control workers performs this task on a daily or monthly basis. They drive or walk to a certain warehouse location, scan the item's barcode, count the units, and then move on to the next scheduled location. This technique has certain disadvantages, despite the fact that it increases inventory accuracy over annual one-time inventory inspections. Cycle counting is a labour-intensive, dangerous (due to high altitude activities), expensive (labour expenditures) and error-prone (among other things) job (highly repetitive tasks). Drones can help make this procedure go more smoothly. Drones for inventory management have three main objectives: increase inventory accuracy, lower labour costs, and minimise the number of potentially dangerous activities for humans.

10.5.4 INTRA-LOGISTICS

Drones have the potential to improve intra logistics. They can, for example, carry parts from warehouses to factory workshops. Drones have a lot of promise for interior applications, such as on-site delivery of tools as well as lubricants and spare parts, because they can follow pre-defined flight patterns and transport items. Payload, gripping/placing movements and navigation are all significant limitations in intra logistics [15].

10.5.5 INSPECTION AND SURVEILLANCE

In warehouses, drones might be a feasible alternative for manual inspection and monitoring activities. Drones are used for inspection in a variety of sectors, including construction, petrochemicals, oil and gas, and power production. Indoor inspections are becoming more common with the usage of drones. Drones may be used to inspect warehouse roofs, racks, pallet placements, walls and ceilings, among other things as shown in Figure 10.4. As warehouse operations and customer demand have risen, inspection procedures have gotten more expensive and sophisticated. Indoor drones are useful for inspection and monitoring activities in potentially dangerous situations or at high altitudes. Drones might potentially be used to follow patrol routes in order to prevent crimes and thefts.

FIGURE 10.4   Ariel view from drone.

10.6   AUTOMATED GUIDED VEHICLES

Self-driving cars are known as AGVs. Around 1954, the first versions of AGVS were introduced. They are frequently employed in flexible production systems, material handling systems and container handling applications to carry material from one point on the plant floor to another without the need for an accompanying human. With the advancement of technology, more complex equipment is now accessible, reducing machining and internal setup time significantly. Along with quick production, the goal of production planning is to ensure effective material transit between workstations and in and out of storage. To conduct an effective routing of material with random handling capabilities, flexible material handling systems are necessary. Because the flow channel may be readily selected from a variety of different paths or changed to suit new sites, the usage of AGVs enhances flexibility. Because it has a direct influence on the complexity of the control system software, installation cost, journey time and the design of the material handling guide route have a major impact on overall system performance and dependability.

10.6.1   ADVANTAGES OF AN AUTOMATED GUIDED VEHICLE SYSTEM

1. Cost of labour is lower.
2. Increased Productivity and Accuracy.
3. A high level of dependability and availability.
4. Programmability allows for random material management.
5. All AGVs are operated in unison.

10.6.2 Different Types of AGVs

10.6.2.1 Fork Lifts

AGVs forklift trucks have just recently become popular. Guided fork trucks are used when the system requires automated pickup and delivery of items from the stand level or floor, and the height of load transfer change at stop locations. Without the requirement for human involvement, the guided fork truck can pick up or discharge freight automatically. The vehicle's prongs may be elevated to one to two metres in height, allowing it to serve conveyors or load platforms of varying heights within a system [25].

10.6.2.2 Tow/Tugger

AGVS towing applications are the original, and they are yet the most popular type of AGV. Large-scale product movement into and out of warehouses, as well as direct service to a manufacturing/assembly plant, are examples of towing applications. Side-route spurs are frequently constructed at receiving or shipping facilities to allow trains to load or unload from the main railway without impeding other trains on the main line. The AGV trains are also commonly utilised to deliver products in the supply chain. In this case, the AGV trains are loaded with products for specific locations along the guided path route.

The train will make many stops in order to unload the cargo at the proper locations. Train systems are frequently used for long-distance product movement, sometimes between buildings, sometimes outdoors and in extremely large scattered systems with extensive routes. This is a very efficient approach that can often be justified merely by removing the requirement for fork trucks, operators and manual trains because each train can transport up to 16 pallet loads at a time [25].

10.6.2.3 Unit Load

Individual load movement is generally assigned to distinct missions in AGV unit load applications. Unit load carriers are commonly used to connect conveyors to storage retrieval systems or assembly/manufacturing processes. They are a very efficient way of horizontal movement between intensive-hardware material handling subsystems in this application. The unit load carrier is capable of transporting large amounts of products over short distances while also connecting other automated subsystems in a fully integrated facility. Unit load systems often include autonomous product pickup and delivery, as well as remote vehicle management.

The unit load carriers are commonly employed in storage and distribution of systems with short guide paths but high volumes. The unit load carriers can move in narrow spaces where AGVS trains would be too cumbersome to utilise. Unit load carriers with lift/lower decks or roller decks make it simple to move loads to conveyors or load platforms. Because they generally function independently of one another and can pass each other to go to specified destinations, unit load carriers provide high system flexibility for product movement [25].

10.6.3 Guidance Methods

10.6.3.1 Wire – Embedded in Floor

A slit is made in the floor, and a wire is placed 1 inch below the surface. This space has been cut out along the AGV's route. This wired cable transmits a radio signal.

On the bottom of the AGV, a sensor is located near the ground. As the radio signal passes down the cable, the sensor evaluates its relative location. The steering circuit is controlled by this data, which forces the AGV to follow the wire [26].

10.6.3.2   Guide Tape (Magnetic or Coloured)

The guiding route for AGVs is made of tape. The AGV is equipped with a guidance sensor that allows it to follow the tape's course. Tape has a number of advantages to cable guidance, including the ability to be readily removed and repositioned if the path has to be changed. Coloured tape is less expensive at first, but it doesn't have the advantage of being able to be utilised in high-traffic locations where it could get damaged or soiled. Although the flexible magnetic bar is buried in the floor like wire, it functions in the same way as magnetic tape and is thus unpowered or passive. Magnetic guide tape also has the benefit of being dual polarised [26].

10.6.3.3   Inertial (Gyroscopic) Navigation

Inertial navigation is another kind of AGV guidance. A computer control system controls and allocates duties to the vehicles using inertial guidance. The floor of the workplace has been fitted with transponders. These transponders are used by the AGV to keep the vehicle on track. A gyroscope can detect even the tiniest changes in the vehicle's orientation and adjust the vehicle's direction to keep the AGV on track. The inertial technique has a one-inch margin of error. Inertial can work in a variety of conditions, including narrow aisles and cold temperatures. The use of magnets installed in the facility's floor that the vehicle can read and follow is one type of inertial navigation [26].

10.6.3.4   Laser – Triangulation from Reflective Target

Reflective tape is placed in buildings, poles, and stationary machinery to aid navigation. A laser transmitter and receiver are placed atop a spinning turret on the AGV. The same sensor transmits and receives the laser. The angle and (sometimes) distance to any reflectors are automatically computed in the line of sight and range. This information is compared to the AGV's memory map of the reflector setup. The AGV's present location may then be triangulated by the navigation system. The current position is compared to the reflector layout map's programmed path.

10.6.4   PATH DECISION

AGVs must make judgements on which course to take. This is achieved in a number of methods, including frequency choose mode (wired navigation only), path choose mode (wireless navigation only) and a magnetic tape on the floor that serves as a guide for the AGV as well as steering and speed commands.

10.6.4.1   Frequency Select Mode

The frequency select mode makes its choice based on the frequencies produced by the floor. When an AGV reaches a location on the wire where the wire divides, it detects the two frequencies and chooses the optimal path using a table stored in its memory. The varied frequencies are only necessary at the AGV's decision point. After this moment, the frequencies might shift back to a single signal. This approach is difficult to scale and necessitates further cutting, which costs more money [27].

10.6.4.2   Path Select Mode

The mode of AGV selects a path from a list of pre-programmed pathways. It compares sensor readings to values programmed into them by programmers. When an AGV comes to a fork in the road, it simply needs to choose between paths 1, 2, 3 and so on. This is an easy decision because it already knows its course according to its programming. Because a team of programmers is required to programme the AGV with the right pathways and alter the paths as needed, this technique can raise the cost of an AGV. This approach is simple to alter and implement [27].

10.6.4.3   Magnetic Tape Mode

The magnetic tape is laid on the floor or buried in a 10 mm channel, and it not only provides a path for the AGV to follow, but it also contains strips of tape with different polarities, sequences, and distances laid along a side the track that tell the AGV to change lanes, slow down, speed up and stop [27].

10.6.5   APPLICATION

In many industrial facilities and warehouses, efficient and cost-effective material transportation is a critical component of optimising operations. Because AGVs may provide efficient and cost-effective material transportation, they can be used in a variety of sectors in standard or customised designs to meet specific needs.

10.6.5.1   Industries Application

1. Manufacturing
2. Chemical industry
3. The pharmaceutical industry
4. Printing and paper
5. Food and drink
6. Hospital
7. Warehousing is a term used to describe the process of storing goods
8. Theme parks

10.6.5.2   Common Applications

AGVs are utilised to carry a range of materials such as pallets, racks, rolls, containers and carts in a number of applications. AGVs are particularly well suited to the following applications:

1. The transportation of materials across a long distance in a repetitive manner
2. Stable loads delivered on a regular basis
3. Medium volume/throughput
4. When punctuality is critical and late deliveries are inefficient
5. At least two shifts are necessary for operations
6. Processes that rely heavily on material tracking

10.7   PLANNING AND CONTROL OF AUTONOMOUS VEHICLES

An autonomous vehicle is the one that operates automatically without the help of any human intervention by identifying its surroundings or environment. Autonomous vehicles can make use of fully automated driving system where it acts like a human driver who can usually manage with the severe damage to its surroundings. Figure 10.5 shows the six different levels of Autonomous vehicles as follows:

- At level 1, human driver must take care of the driving, thus the car doesn't hold any control in its operations.
- At level 2, Advanced Driver Assistance System (ADAS) in vehicles supports any among the following such as steering or braking or accelerating.
- At level 3, ADAS can perform steering or braking or accelerating, but still the human driver is required to manage the overall performance of the car though ADAS system performs everything.
- At level 4, ADAS system performs all the control over the car operations, but it still requires human driver then and when ADAS system is requested.
- At level 5, human intervention is not required and so ADAS system could perform on its operations by its own.
- At level 6, the driverless car gets operated on its own without involvement of any human drivers. Also it makes use of 5G technology where it enables the driverless cars to communicate with the other driverless cars, following the traffic signals perfectly and as well as the road signs are sensed.

Some of the advantages of the autonomous vehicles include the following: (1) Since these vehicles follow traffic signals, road signals and other road safety measures, they decrease the impact of road accidents which in turn leads to less death rate when compared to human driving cars. (2) Also they decrease the traffic crowding. These traffic congestions could be sensed by the means of highly automated vehicles.

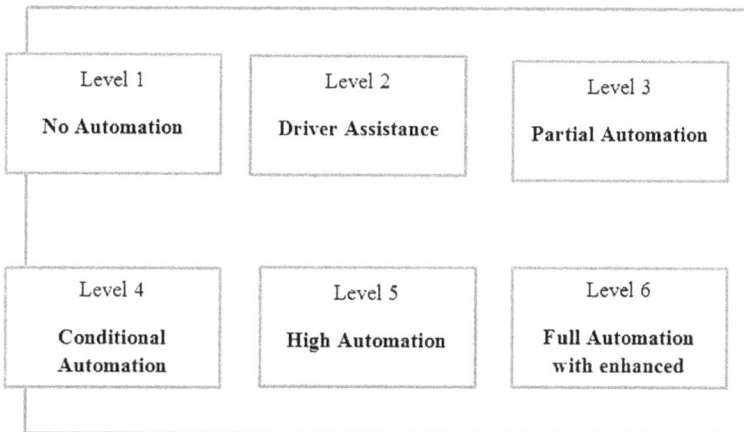

FIGURE 10.5   Levels of autonomous vehicles.

(3) They also ease the elder people with these driverless cars. In order to enhance the efficiency of the ADAS system, lots of time and human efforts are put forward in terms of technology, thus increasing the efficiency of the system.

To make things done effectively, decision making and planning is needed in the following points: (i) to ensure from where does the journey get started and where the destination must end with; (ii) to choose the best optimal and the shortest path; (iii) the way interactions are performed with other driverless vehicles; (iv) the learning way they learn to drive from the human driving history and (v) to interact with other vehicles, peoples and any other obstacles such a way that it reaches the destination in a safer way.

Path tracing is the important ability of an autonomous vehicle and which targets to complete the planned tasks safely and more effectively. The wheeled mobile robot is enabled with the torque controller for tracking the scheduled path. The backstepping control algorithms are implemented in the autonomous vehicles for tracing the path on the curve road. Based on the lateral dynamic model of the vehicle, sliding mode variable structure path tracing regulator has been used to design the radical basis function neural networks. This design helps the automated vehicle to plan the virtual path among the automobile mass centre and forecast aiming point according to the vehicles kinematic model [7].

The longitudinal and lateral control with the emergency braking system based on the control theoretical methods have been computed to develop an autonomous vehicle which is probably free from the collisions. The communication between the autonomous vehicles can be exchanged with the help of on-board computation and wireless technologies. The information related to the road infrastructure, traffic data, closely related areas will be communicated between the two autonomous vehicles which are moving the particular city or area [10].

The autonomous vehicles contain the preceptors to gather the information and to mine the significant knowledge from the environment. The planning software used to predict the focused decision to make the robots to think in higher orders. The actions are planned and executed by the control part [5]. Figure 10.6 shows the outline of autonomous vehicle with its core competencies.

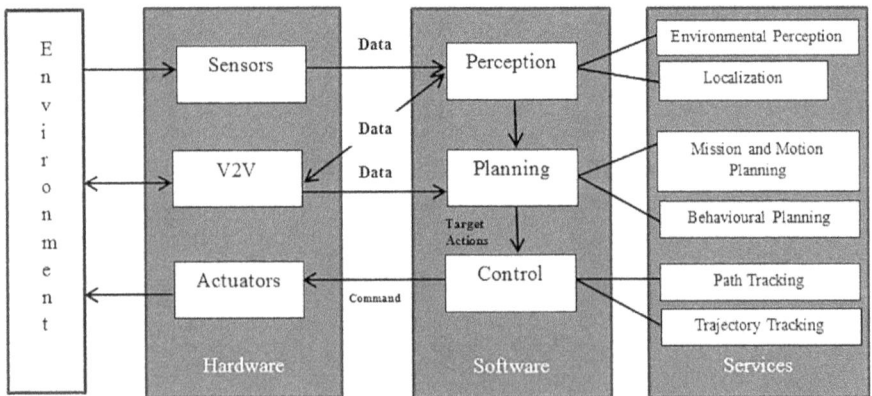

FIGURE 10.6 Overview of autonomous vehicle system.

10.7.1 INTEGRATED PLANNING

Automated vehicle need to detect the right path way to reach the destination in the local map, though number of salient features are included in it, there are some disadvantages leading to fewer mistakes. Because it is highly not possible to take a change in route like how a human driver does when they come to know the misleading path or any work behind that pathway quickly. Whereas the highly effective maps are much costly to create and maintain because those data need to be consistently monitored and updated. Though the light-weighted localised maps have disadvantages, it is to be noted that they are good to be provided in the autonomous vehicles rather than the highly effective maps because of its maintenance.

In the real world, robots that are employed with intuitive models must be able to properly handle situations in which they are required to make decisions in conditions that are not similar to those in their training. Recent research shows that some methods like ensemble, bootstrap for measuring the neural network indecision may not provide the perfect uncertainty result unless they are provided with same training dataset. Other set of researchers found that auto-encoders seem to provide the best results in which they could perform by learning from their mistakes. Thus, they observe a lot of failure cases and perform their operation [28]. It is also to be noted that this reinforcement learning works better with the best case scenarios rather than the worst case where it is supposed to come across any road crashes. To come across all sort of situations well-designed and well-planned strategies are implemented.

10.7.2 SYSTEM ARCHITECTURE OF PLANNING AND DECISION MAKING IN AUTONOMOUS VEHICLES

Planning and decision making in an autonomous vehicle shown in Figure 10.7 involves the following layers:

(i) **Sensor Layers:** These layers are fitted in such a way that they hold the camera by which they can sense the complete real-time environment. Also they are used to detect the vehicles and obstacles around them. Incorporated vehicle navigation locates the car and also gets its current information.
(ii) **Mission layers:** These layers provide the complete mission, traffic rules and road signs to be followed correctly.
(iii) **Perception layers:** These layers include maximum functionalities like identifying obstacles, road edges, traffic signs as well as vehicle detection.
(iv) **Geographical Information System (GIS) layer:** It includes providing data upon global path planning and road networks.
(v) **Path planning:** This layer is required to identify the right pathway by avoiding the obstacles in the road path and also by obeying the traffic rules.

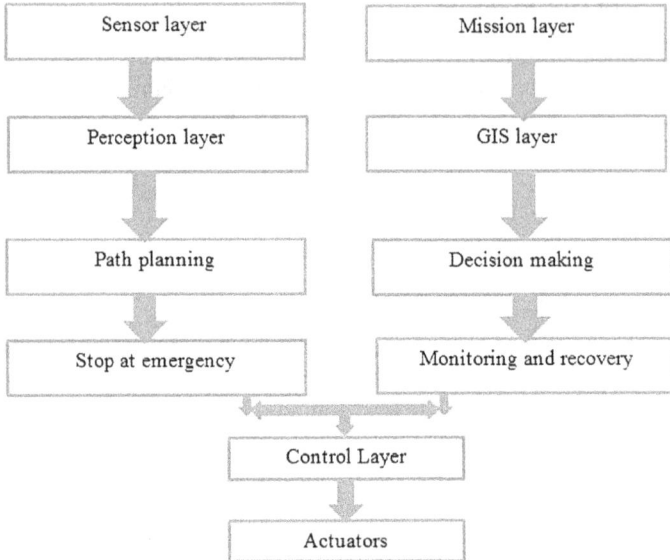

FIGURE 10.7 Planning and decision making system architecture.

Overall system architecture for autonomous vehicle's planning and decision making systems is provided below:

(vi) **Decision making layer:** This layer is helpful in making the decision by obtaining information from the other layers to make the next move and also to calculate whether or not to go in high speed in that specific environment.

(vii) **Control Layer:** This is the one that acts as entire control system of the car. The car need to follow the right path with information from the path planning layer and making the next move with decision making layer. Combined with these layer strategies, this layer is responsible to control the car parts like steering, gearing, accelerator and braking systems.

(viii) **Actuator layer:** Actuators are responsible for moving and controlling the overall mechanism of a car. Actuator in a car includes motor, lights, etc..

(ix) **Stop at emergency layer:** In case to face with car's safety at some situations like accidents, emergency stop layer is widely used.

(x) **Monitoring and recovery layer:** In this layer, the maintenance phase is followed. It means that when any part of the car is repaired, then it is recovered then and there with the help of this layer.

With the help of these layers, much functionality is done. Layers that play an important role in this system include a control layer which controls the overall mechanism of the car, decision layer to pick the right decision to keep forwarding with the next move. Also the actuator layers are responsible for moving and controlling the car. Important functionalities of path planning and decision-making system include generating the real-time tracking of the car and also guaranteeing with the path planning. Main idea behind the path planning algorithm is to choose the best optimal path from the existing parallel lines to provide it with minimal cost efficiency [29].

10.7.3 DEEP NEURAL NETWORKS (DNN) IN OBJECT DETECTION AND CLASSIFICATION

DNN detects impending obstacles, traffic lights and road signs. LightNet identifies traffic light whether red, yellow or green. Though DriveNet identifies other vehicles, traffic lights and road signs, they do not read light colour. WaitNet identifies the condition upon where the vehicles must stop and wait depending on the other vehicles on road. SignNet identifies the road signs. Three primary objects used in autonomous vehicles include Camera, Radar and Lidar. Cameras are used for taking photos to the nearby vehicles, or any other objects. An autonomous vehicle uses camera data to perceive objects in the environment. These data were further used for parking system. Numerous computational related algorithms have been discussed for various applications by sathish and karthiga et al. [3,27,30,31]. Radars are used in delivering the highly accurate radar sensing technology to detect the other cars and perceive its environment. Lidar is used in autonomous vehicles provided with 3D technology just to identify the global path planning with geographical road pathway.

10.8 CONCLUSION

In recent years, the complications and necessities in the automobile industries have steadily raised due to the factors like increasing market volatility, international market competitions and demand for the personalised products. To regulate the prevailing demands, many automobile industries tend towards the automation to enhance the flexibility, sustainability, stability, adaptability and to be as a competitor in the global markets. The digitisation in the automobile industries laid the path for AMRs, AGVs, cobots and so on. The assimilation of cyber-physical systems with IoT in the logistics facilitates material tracking and transporting without any precise risk. The possible use of AGV, AMR in the automotive industries and their routines for smart engineering have been deliberated thoroughly. The planning and decision making by the autonomous vehicles have been considered for mainly identifying the correct pathway without any object collisions. Thus, the automation in the vehicle industry brings the numerous advantages to the human society such as quality product on time, no harm to environment and improved safety and secured products.

REFERENCES

[1] Wang, H., Wang, B., Liu, B., Meng, X., & Yang, G. (2017). Pedestrian recognition and tracking using 3D LiDAR for autonomous vehicle. *Robotics and Autonomous Systems*, 88, 71–78.
[2] Felzenszwalb, P. F., Girshick, R. B., McAllester, D., & Ramanan, D. (2010). Object detection with discriminatively trained part based models. *IEEE Transactions on Pattern Analysis and Machine Intelligence*, 32(9), 1627–1645.

[3] Zhang, Biao, Martinez, C., Wang, Jianjun, Fuhlbrigge, T., Eakins, W., & Chen, Heping (2010). The challenges of integrating an industrial robot on a mobile platform. In *IEEE International Conference on Automation and Logistics (ICAL)*, 255–260.

[4] Amiruddin, B. P., & Romdhony, D. R. (2020). A study on application of automation technology in logistics and its effect on E-commerce. https://doi.org/10.31224/osf.io/vs9yg

[5] Son, D. W., Chang, Y. S., Kim, N. U., & Kim, W. R. (2016). Design of warehouse control system for automated warehouse environment. In *5th IIAI International Congress on Advanced Applied Informatics (IIAI-AAI)*, Kumamoto, Japan, 980–984. https://doi.org/10.1109/IIAI-AAI.2016.188

[6] Salkar, C. M., Tapare, G. J., Murkute, M. A., Zingre, C. R., Mohod, H. A., & Gupta, V. S. (2021). Field Data Analysis Using Work Measurement Techniques in a Packaging Industry. In *Advances in Mechanical Engineering: Lecture Notes in Mechanical Engineering* (pp. 371–376), Springer, Singapore. https://doi.org/10.1007/978-981-15-3639-7_43

[7] Thomas Renald, C. J., Yuvaraj, S., & David Rathnaraj, J. (2021). Cell Arrangement Measurement—A Perfect Usage of Cell Fabricating Framework in an Industry. In *Innovative Design, Analysis and Development Practices in Aerospace and Automotive Engineering: Lecture Notes in Mechanical Engineering* (pp. 11–19). Springer, Singapore. https://doi.org/10.1007/978-981-15-6619-6_2

[8] Joshi, D., Nepal, B., Singh Rathore, A. P., & Sharma, D. (2013). On supply chain competitiveness of Indian automotive component manufacturing industry. *International Journal of Production Economics*, 143, 151–161.

[9] Bagalagel, S., & ElMaraghy, W. (2020). Product mix optimization model for an Industry 4.0- enabled manufacturing-remanufacturing system. *Procedia CIRP*, 93, 204–209. https://doi.org/10.1016/j.procir.2020.03.029

[10] Ebrahimi, M., Baboli, A., & Rother, E. (2019). The evolution of world class manufacturing toward Industry 4.0: A case study in the automotive industry. *IFAC-Papers OnLine*, 52 (10), 188–194. https://doi.org/10.1016/j.ifacol.2019.10.021.

[11] Liu, J.-L., Wang, L.-C., & Chu, P.-C. (2019). Development of a cloud-based advanced planning and scheduling system for automotive parts manufacturing industry. *Procedia Manufacturing*, 38, 1532–1539. https://doi.org/10.1016/j.promfg.2020.01.133

[12] Hofmann, E., & Rüsch, M. (2017). Industry 4.0 and the current status as well as future prospects on logistics. *Computers in Industry*, 89, 23–34.

[13] Alshamsi, A., & Diabat, A. (2015). A reverse logistics network design. *Journal of Manufacturing Systems*, 37(3), 589–598.

[14] Oyekanlu, E. A., et al. (2020). A review of recent advances in automated guided vehicle technologies: Integration challenges and research areas for 5G-based smart manufacturing applications. *IEEE Access*, 8, 202312–202353. https://doi.org/10.1109/ACCESS.2020.3035729.

[15] George, D. (2012). Why should we use autonomous industrial mobile manipulators – Smashing robotics. https://www.smashingrobotics.com/why-should-we-use-autonomous-industrial-mobile-manipulators/. Accessed 13 August 2021.

[16] Groover, M. P. (2020). Automation. Encyclopedia britannica. https://www.britannica.com/technology/automation. Accessed 12 August 2021.

[17] Cary, H. B., & Helzer, S. C. (2005). *Modern Welding Technology*. Pearson Education, Upper Saddle River, NJ, p. 316. ISBN 0-13-113029-3.

[18] Turek, F. D. (2011). Machine vision fundamentals, how to make robots 'see'. *NASA Tech Briefs Magazine*, 35(6), 60–62.

[19] Aroon Dass, P., Rakesh, S. (2018). Automated guided vehicle system. *International Journal of Engineering Research & Technology, Confcall*, 6 (14), 1–4.

[20] Buniyamin, N., Wan Ngah, W. A. J., Sariff, N., Mohamad, Z. (2011). A simple local path planning algorithm for autonomous mobile robots. *International Journal of Systems Applications, Engineering & Development*, 5(2), 151–159.

[21] Lichao, X., Kamat, V. R., & Menassa, C. C. (2018). Automatic extraction of 1D barcodes from video scans for drone-assisted inventory management in warehousing applications. *International Journal of Logistics Research and Applications*, 21(3), 243–258. https://doi.org/10.1080/13675567.2017.1393505

[22] Companik, E., Gravier, M. J., & Farris, M. T. (2018). Feasibility of warehouse drone adoption and implementation. *Journal of Transportation Management*, 28(2), 31–48. https://doi.org/10.22237/jotm/1541030640

[23] Alias, C., Salewski, U., Ortiz Ruiz, V. E., Alarcón Olalla, F. E., Neirão Reymão, J. D. E., & Noche, B. (2017). Adapting warehouse management systems to the requirements of the evolving era of industry 4.0. In *Proceedings of the ASME 2017 12th International Manufacturing Science and Engineering Conference.* https://doi.org/10.1115/MSEC2017-2611

[24] Rajesh Kanna, P., & Vikram, R. (2020). Agricultural robot—A pesticide spraying device. *International Journal of Future Generation Communication and Networking*, 13(1), 150–160.

[25] Mueller, T. (1987). *Automated Guided Vehicles*, edited by R. H. Hollier (p. 277). IFS Publications/Springer Verlag.

[26] Premi, S. K., & Besant, C. (1983). A review of various vehicle guidance techniques that can be used by mobile robots or AGV's. In *Proceedings of the Second International Conference on AGV Systems.* IFS Publications Ltd, UK.

[27] Yu, Y., Wang, X., Zhong, R. Y., & Huang, G. Q. (2016). E-commerce logistics in supply chain management: Practice perspective. *Procedia CIRP*, 52, 179–185. https://doi.org/10.1016/j.procir.2016.08.002

[28] Schwarting, W., Alonso-Mora, J., & Rus, D. (2018). Planning and decision making for autonomous vehicles. *Annual Review of Control, Robotics, and Autonomous Systems*, 1, 187–210.

[29] Mengyin, F., Song, W., Yang, Y., & Wang, M. (2015). Path planning and decision making for autonomous vehicle in urban environment. In *IEEE 18th International Conference on Intelligent Transportation Systems*, 686–692. https://doi.org/10.1109/ITSC.2015.117

[30] Karthiga, M., Sountharrajan, S., Nandhini, S. S., & Kumar, B. S. (2020, May). Machine learning based diagnosis of Alzheimer's disease. In *International Conference on Image Processing and Capsule Networks* (pp. 607–619). Springer, Cham.

[31] Sathish Kumar, G., Premalatha, K.(2021), Securing private information by data perturbation using statistical transformation with three dimensional shearing. *Applied Soft Computing*, 112. https://doi.org/10.1016/j.asoc.2021.107819.

11 Drones in Agriculture
Multispectral Analysis

Abhishek Choubey and Bharath Chandan Reddy

Sreenidhi Institute of Science & Technology, Hyderabad, India

CONTENTS

DOI: 10.1201/9781003181668-11

11.1 INTRODUCTION

The revolution in aerial technology especially UAVs (Unmanned Aerial Vehicle) [2] is widely used in different sectors. One of the most important features is that it can be utilized to collect the images in high resolution from the top view and helps in taking further action. Considering the rise in environmental changes, human population, and land availability, technological innovations are required to increase productivity in agriculture. With the improvement, remote sensing sensors integrated into UAVs, which helped in collecting huge data that became a tool that has opened up for logical exploration in the field of precision agriculture [1]. From the previous decade, research has been expanded with the consequence of the increase in the agriculture image data prompted advancements in evolving precision agriculture. The multispectral analysis of aerial data is obtained by capturing the images from the multispectral camera sensor that is integrated with the Unmanned Aerial Vehicles (UAVs) that are used in applications determining the crop health prediction, analysis, and generating the reports with recommendations to take the action within the field precisely. The multispectral analysis helps in detecting the crop health of the farm and gives the results to the farmers, researchers, and agronomists to proceed to the next step. Multispectral analysis helps in the entire crop cycle, i.e. from leveling of land to crop harvesting.

Traditionally, Agriculture data has come from scouting in the farm field. But the cost, judgment subjectivity, and occasional errors in human scouting limit its effectiveness compared to newer alternatives due to technology development. The new measurement technologies are expanding our ability to understand in estimating the crop performance considering all the factors that impact the productivity. In situ sensors can now measure soil moisture, micro, and macronutrients where the mounted sensors can measure spatially variable yield. Aerial and satellite remote sensing systems offer a range of new metrics to predict productivity through various plant indices. There is a necessity of integrating data accumulated from various sources to extract the most application-oriented data. It can very well be presumed that aerial data has a few difficulties to make them into a useful one. These remember a better understanding of the variables influencing the association between electromagnetic radiation and vegetation in a specific climate, thereby choosing proper spatial and spectral data for handling strategies to extract image of the farm area.

Using remote sensing technology, crops are monitored throughout the season with regular aerial imagery at key decision times. Along with nitrogen application rates that are recommended for each zone based on estimated yield response, researchers also developed a systematic approach for detecting weeds, pest attack, fungal infection, and water draining models. It is helping the farmers to make the decision quickly that helps in reducing the damages. Crop comparison in each different season at the right time applies machine learning to remote sensing imagery and user observation to identify crop stress in the field, estimate its extent, and promptly alert the farmer.

11.1.1 Drones

UAVs integrated with the multi-spectral camera are used in precision agriculture for several applications, including capturing the image for analysis of crops in the farm, detecting a deficiency of macro and micronutrients of the crops, acquiring data on soil water holding capacity or the executive's water system frameworks particularly for enormous agricultural producers that develop in locales with scattered zones. In the plan, improvement and execution of UAV frameworks, coordinated with designing factors like aerodynamics, hardware, type of materials, PC programming. Presently, UAVs are additionally being utilized in different industrial applications. such as mining, industrial inspection, security surveillance, and solar inspection. These include the utilization of UAVs such as quadcopter or fixed wings in precision farming for pesticide spraying and multispectral analysis [3].

In general, agricultural land covers huge areas with a single crop. Using the quadcopters requires a huge investment and covers less land with the low flight time. Using fixed-wing drones overcomes the disadvantages of quadcopters.

11.1.2 Multispectral Camera

The multispectral camera is a sensor that was designed specifically to integrate with the drone for easy usage. It captures non-visible data of crops over five distinct spectral bands: blue, green, red, red-edge, and near-infrared (NIR). The multispectral analysis involves extracting images, image optimization georeferencing, calibration, and Ortho mosaics. The accurate maps allow the agronomists, researchers, and farmers to give a precise crop assessment for every square meter of a farmer's field. It helps in assessing vegetative health mapping, disease detection, weed detection, plant count and classification, detect and monitor deficiencies, terrain modeling, and drainage facilities [4]. It also helps in insurance estimating the crop failure due to natural and other reasons. Apply the advanced scientifically validated analytics to create nutrient content-management zones. In terms of field coverage, the drone's land coverage varies and depends on the weather, required pixel per cm, the field of view, and the type of drone used in acquiring farm data.

The data collected from the camera is stored in the SD card and it can be transferred to computers by transferring wirelessly or connecting the SD card to the computers. The images processed using multispectral analysis through cloud-based analytics or system software. The output is generated from the analytics software along with the recommendations and predictions.

The multispectral camera in general requires a trigger input from supporting hardware. Recently, few multispectral cameras are internally triggered based on the input from the altitude. The only required connection is power in general given from the battery connected to UAVs. The trigger rate can be configured using the mapping software after selecting the overlapping percentage, flight time, land coverage, and type of UAV. The multispectral camera sensor, wavelength and bandwidth details are presented in Table 11.1.

TABLE 11.1

Multispectral Camera Sensor, Wavelength and Bandwidth Details

Multispectral Camera Sensor	Mid-range Wavelength	Bandwidth
Blue	475	32
Green	560	27
Red	668	16
Red edge	717	12
NIR	842	57

Image calibration helps to collect accurate, iterative data in the different lighting conditions. It presents to show the different sets of data over time during the crop cycle and the capability of drone imagery obtained from the multispectral camera [5].

The image data collected from the camera gives an assortment of fair choices for both the image data and the plant indices layer to permit you to utilize the information in the software management system. The multispectral image data and software are firmly coordinated to access the unique capacities to get the benefits for using this technology, for example, conquering the mistakes because of cloud, non-covering, and shadows of UAV and generate the plant index layers. To build the land inclusion for a given flight time, scientists are chipping away at photogrammetric procedures where we can separate the subtleties with simply a cover of images, which expands the efficiencies of UAVs. Overlapping the images need to be expanded during unfavorable conditions such as cloudy weather and certain applications, yet under 70% cover ought to be adequate for most flights.

When the images from drones are obtained, the image processing starts on the actual sensor during its flight. The plant indices are obtained finally on the surface once multispectral image data is collected, the time taken relies upon the area of farm and amount of data collected and the hardware used for the processing of the image data.

Later once the data is uploaded to the cloud, it has a set of advanced image processed algorithms in providing accurate, actionable, information to growers, agronomists, and researchers, more than just aerial images. The data generated from the analysis of multispectral image data which includes high-resolution images along with the geotag, generated the layers of plant indices such as NDRE (Normalized Difference Red-Edge), NDVI (Normalized Difference Vegetation Index), OSAVI (Optimized Soil-Adjusted Vegetation Index), Chlorophyll map, and User-Defined Smart Detection. Many new algorithms are being developed on the type of crops all over the world. Processing the algorithms required a minimal hardware requirement. Each map obtained is geo-tagged and the individual images and the image processed maps can be exported as KML (Google Earth), SHP (Shapefile), and Geo-TIFF. All the data products can be exported as a shapefile, which is the file format used in farm management software [6].

It's easy to upload images from the SD card in Geo-TIFF format and these geo-tiff layers can be also used in GIS software like ArcGIS, QGIS, and Google Earth.

However, processing provides added benefits, estimating the front and side overlap, features offline and online processing, and using the data for customized algorithms. We can add customized functionality to obtained results [7].

11.1.2.1 Capturing the Multispectral Data

Among alternatives, low-altitude spectral imaging systems offer unique advantages. The combination of high spatial and spectral resolution enables new imagery intelligence techniques that measure plant morphology, chemical and physical composition throughout various stages of maturity. Best practices for capturing quality data. Fly within 2 hours in the afternoon by ensuring camera points tilting with deviation in ten degrees angle, with accurate GPS (Global Positioning System) for geotagging images, and ensuring side lap and front lap overlap with specified percentage, and capture reflectance panel image immediately before and after each flight to calibrate the images, without impacting the images with shadows. In general, the flight plan is selected by the drone that depends on the required resolution, flight time, area of coverage, and flight time of the drone. While planning your mission, it is important to maintain the distance between tracks. Novel geospatial intelligence methods provide a broader range of insights with efficient and scalable solutions.

By using the Reflectance Panel [8], it ensures that the image quality does not go down due to shadows, since during the flight the images might capture the images along with the shadows of drones. The DLS attached to the drone must be parallel to the ground to calibrate the images and remove the shadows from the images. The reflectance panel helps to calibrate the multispectral data [5]. The multispectral camera [9] details are represented in Table 11.2.

11.1.3 DISCUSSION (LITERATURE SURVEY)

11.1.3.1 Difference between Multispectral Data Obtained from Satellite and Drone

Satellites and UAVs multispectral cameras are not contradicting advancements; indeed, the two wellsprings of information utilized for the comparative applications and their data, in general, will be complementary with minute differences. For example, integrating the satellite and drone imager data provides in-detailed data and varied information for the specialists and customers.

TABLE 11.2
Multispectral Camera Details

Camera Name	Mutlispectral Camera
Focal length [mm]	5.4
Sensor width [mm]	4.8
Image width [pixel]	1280
Image height [pixel]	960
Min. triggering interval	1

The satellites and UAVs are integrated along with multispectral cameras that are used in remote sensing applications. Natural vegetation present on the surface of the earth absorbs and reflects light received from the sun. The reflection of light is through image sensors captured by satellite and aerial robots, which can be present in the visible spectrum or other frequencies. Natural eyes are intended to identify the only visible spectrum of light; In any case, multispectral cameras and satellites can detect and quantify invisible bands of the electromagnetic spectrum that gives the data that helps in developing new algorithms in detecting the abnormalities of the crop and help in taking the decision particularly for the farmers and researchers all over the world.

Satellites detect the surface of the earth with their sensors and give information on changes after some time. Multispectral camera sensors are utilized to detect the entire planets alongside explicit chunks of land and give information on how those territories change over the long run. The information gathered with the two innovative technologies is usually utilized for the detection, classification, and planning of vegetation, as a system of being less expensive and less tedious than manual surveys in the farm field [10–12].

Photogrammetry software measures and analyzes the satellite multispectral image data to give exact per-tree analytics of health and feature trouble spots that could somehow or another go undetected [13]. From this plant index layer, farmers can monitor failing trees and take actionable decisions on their farm field precisely. With more than 10 bands, multispectral sensors capture image data from frequency ranges across the visible spectrum of light, typically including green, red, blue, red edge, and NIR. The band within a sensor decides its uses and applications which are generally identified with changes in land use, Plant vegetation mapping, detecting and scanning the presence of natural resources, others [14].

The Multispectral camera, for instance, uses five bands of red, green, blue, NIR, and red edge to gather image data utilized in applications like vegetation mapping of the crop, disease detection, crop stress detection, water stress, and management of irrigation. These abilities are not restricted to UAVs; even a few satellites additionally use these multispectral sensors to capture image data from the surface of the earth.

Satellites are utilized to screen the earth and gather data on various natural phenomena happening on the earth's surface. These particular applications generally depend upon depending on the sensor that is utilized and the height of the satellite orbit from the surface. There are notable satellite projects in the remote sensing areas and their noteworthy image data is openly available for analysts and researchers all around the world.

The image data captured by satellite is used to scan the changes on the surface of the earth and the relationship of those progressions' different events like environmental and weather change, urbanization, and rapidly spreading fires in the forests. In general, the satellite using the camera sensors more than 13 frequencies of the visible spectrum, in addition to NIR and shortwave infrared, provides image data used in plant vegetation mapping, soil and water scanning, and detection and monitoring of coastal regions and waterways.

The primary contrasts between satellites and multispectral cameras lay in the expense, the feasibility of the data, the resolution of the spectral and spatial image data, and having the control of data like atmospheric phenomena like clouds. Historic satellite data is accessible at no expense, helping the researchers to compare changes in the earth's surface for a long time free of cost. Notwithstanding, in places where the weather conditions limit the days without the clouds, satellite data might be unclear and fragmented.

The resolution of spatial data from satellites can likewise be an issue on specific applications, similar to plant phenotyping. The data from satellites have a ground resolution less than 10m/pixel, while multispectral cameras have a resolution higher than 10cm/pixel. The spectral data from UAVs have a higher resolution but are more costly per square meter compared with the satellite data.

The inclinations on others depend on the applications for which the image data is required, yet that does not imply that each source can be utilized in turn since satellite and aerial image data can be integrated. Satellite data has an extended notable record than drone-based sensor image data in terms of utilizing for a long time and the bands captured by both systems are different, few issues can be more evident with satellite image data than UAVs and vice versa.

Agronomists all around the world are using both satellite and multispectral image data to give various levels and sorts of data for calculating the vegetative indices. The UAV's spectral data is reciprocal to satellite image data, which we actually use for the overall observation of our crop harvests consistently throughout the year, alongside the software supporting tool. The resolution of multispectral data from the drone is quite helpful in determining the accuracy of plant indices compared to the data obtained from the satellite. Indeed, for giving recommendations related to the treatment on spraying pesticides, adding fertilizers, the robot is best adjusted to the work because of the resolution of the image data. By utilizing the aerial robots for harvest exploring, integrated with plant indices layers and analysis, in addition to our correlative controls, it can furnish fertilizer treatment guidance by two days after a flight [15].

11.1.3.2 Using Data with Plant Indices

The improvement of techniques that can precisely identify physiological pressure in agriculture fields brought about by organic and inorganic factors is imperative for guaranteeing a good product that can fulfill the need for the increase in population. The development of new sensors and stages presents options to increase and customize the practical solution by integrating distantly detected crop data into practical data for using them in agriculture fields. Tried the affectability of multispectral crop data from time-based automated UAV system and satellite data helps to distinguish the type of plant stress in the agriculture farm. The outcomes from both information sources uncovered that both information sources were touchy to physiological pressure in the agriculture field. The drone integrated with multispectral sensor-generated data was more delicate to changes at a better spatial goal and could distinguish pressure down to the degree of individual homestead fields. The satellite image data tried could just distinguish physiological crop stress in a group of at least four trees. Resampling the drone image data to a similar spatial goal as the satellite symbolism uncovered that the distinctions in affectability were not exclusively the aftereffect of spatial goal. All things are considered; vegetation plant indices fit to the sensor attributes of every stage that were needed to enhance the location of physiological stress from the multispectral data of the crop in the field. Our outcomes characterize both the spatial saturated point and the standardized vegetation lists needed to execute observation within the field. An examination between time-arrangement datasets of various sources showed that the two sensors are viable and can be utilized to convey an improved technique for observing the physiological stress of the crop. We tracked down that the higher goal drone data was more delicate to fine-scale examples of herbicide incited physiological stress than the RGB camera sensor. Albeit satellite image data is less delicate for a small area, it is more comprehensive when used for larger areas.

The assistance of agronomy specialists who relate the plant vegetation indices with integrated information like soil, climate, pests, etc. can help the growers increment their crop yields and improve the nature of their harvests. By examining image data from multispectral analysis with an agronomic methodology in building the ecosystem that goes far above NDVI maps empowers us to give dependable counsel to enhance the utilization of nitrogen, identifying weeds and spraying of pesticides precisely. It is tied in with applying the precise measure of nitrogen, at the correct time, in the correct field where the crop stress is detected. In conclusion, alongside the utilization of plant indices from spectral data got from the multispectral analysis, agronomist groups utilize the guides delivered to improve the information for the farmer about their fields and thus better prompt them. The outcome is at least a 10% hike in crop yields.

Every crop yield has various stages during its cycles in a given season, the UAV's multispectral image data is utilized from the starting phase from draining facility inside the field to nitrogen ingestion at important phases of the crop and predicting the production at the final stage of harvesting. Technological innovation should bring the farmers looking exclusively to improve the productivity than before. A flight happens during the various stages to increase the crop quality—for example, increment the measure of protein in the crop—the drone must fly at the particular stages. Therefore, farmers must find the important stages of the crop, when fertilizers can be added, and spraying of pesticides. In general, drone flight is then arranged about seven days before nitrogen application, and the information is used to fluidly apply the season's variable application of fertilizers. This helps a great deal about the size and strength of the harvest, just as the measure of nitrogen that has not been consumed since the farmers' application during his past; the point is to prompt a farmer to apply the fertilizer in precise quantity.

The technological innovation can energize those that are unfamiliar with it and are interested about its abilities, additionally comprehended that to be fruitful within the farm sector you are expected to comprehend what are the grower's requirements and recommendations using this technology. It was initially restricted to get an appropriation since it takes time to get adapted to the new technology and its implementation in the field.

The framework was by and large the thing longed for. Subsequent to taking a gander at a few choices, the perceived technological innovation would give more resolution of the image data, fast in-field processing of data management system, and calibrated aerial data. Growers have seen all the variations in the kind of image data, satellite for instance, and they become tired of another mechanical innovation of technology in their agriculture farm fields since they have not given a satisfactory measure of significant worth heretofore. The way had the option to effectively change the thought of working with drones into an adaptable platform and by deciphering the image data they gathered into a justifiable farm solution for the farmers they were working on precisely.

The analysis of multispectral image data involves using the high-resolution data to process plant indices with point-to-point information from the shape of the field, crop data with the geotagged information. In the process of creating an ecosystem, it is possible to integrate this data into the agriculture machinery. Utilizing this cycle makes records of fluctuating goals on a case-by-case basis by UAV sprayers, farm machinery, weed removing robots, and whatever else the producers can use in the precision agriculture management system.

Presently, farmers have many hectares of different varieties of crops grown in their fields. The most widely recognized crop that is grown in a country like India is Paddy. Farmers are growing the cotton in the huge farmland after the crop paddy. We get precise results when the drone flies over the crops in the row and columns which helps in assessing the image data and building plant vegetation indices layers and determine which parts of the cotton field have been affected by pests and take suitable action.

Farmers discover esteem in this innovation when they were drawing in with high-esteem crops with good productivity in the locale. They realized that doing things another way could at least change the output by 10–15% on average, yet they didn't know about what was accessible to them previously. Robot pilots can design trips with ranchers the day preceding an agronomist will be nearby so they can plan the field and rapidly measure the information before their appearance. Toward the back, they would utilize the innovation in a tweaked approach to suit their requirements for variable-rate applications. It was truly incorporating the information into their work process such that farmers could see beyond the present technology and manage to adapt to a customizable design, so they are having the option to truly deal with this information in their own particular manner and motivation. It assists with filling the gap and gives a productive result. Utilizing the multispectral data gathered using UAVs, image processing software prescribes a variable rate solution to apply a precise measure by recommendation in the generated final report. Also, it saves the producer on input expenses and ensures that the perfect measure of pesticides is applied into areas of the farm field.

We can tell a great deal from taking a gander at your field from the drone collected data from the top view, yet the genuine estimation of elevated information comes from making a move in the field. Giving farmers crop information and showing them pain points in their fields to start work with UAVs crop yield information gathered from multispectral sensors. There are numerous approaches to take action with the data collected. Making an interpretation of this data into noteworthy insights for the grower is the way to convey them in utilizing new innovating technology.

In order to recommend the farm to agronomists and farmers encourage a completely utilitarian administration choice interaction which gives high-resolution plant vegetation maps while utilizing multispectral sensor innovation to help in testing the crop with high resolution of data with advanced image processing algorithms empowers this significant data to be advanced and customized shapefiles of each chunk of the farm field. With help of fixed-wing drones can collect the spectral data of 1000 acres. It gathers the data from the field about the crop and the reports with the spectral that can see the changeability of yields inside the field. Though the farmers were interested to see the report since it's the first time, they can see the data which cannot be seen by the naked eye. Yet they battled to set the information into actionable in the field.

The solid foundations in flight and aviation flying needed to utilize encounters and this innovation as a device to collect the image data. When the flight began, visiting the field along with the farmers and discover the issues they are facing in the field and develop a methodology to help them by using the multispectral data. Rather than exclusively depending on the innovation they needed to think about the reasonable components of cultivating crops during the entire cycle. It is the obligation to comprehend that farmers are cultivating for ages without any aerial technology, so giving valuable data in the form of reports with recommendations can help farmers understand and apply in their field.

There was somewhat of an obstruction in the work process at the ground level that while the farmers could connect with it as far as being focused on testing they couldn't follow up on the information. They couldn't place it into their homestead the executives programming effectively, relegate zones and move it over to a regulator for a variable rate application as per the requirements of the yield. While innovation can energize those that are unfamiliar with it and are interested in its capacities, likewise, comprehended that to be effective inside the farm field area you expected to comprehend the rancher's requirements for an increase in productivity. This is the reply we receive in the initial stage and it was a restricting variable to get appropriation since it's alright to disclose to them it's an issue however assuming you can't fix that issue initially since there is no ideal situation.

11.1.3.3 Farmers Need Solutions from Data

The estimation of multispectral information was not resolved to be totally conservative or time-sensitive yet rather a mix of both of them. Stressing the work process that is made by consolidating high-goal multispectral image data and variable rate application permitted to farmers to see the extended picture of using technology, describing this work process when it is perceived by the farmers, will assist them with changing how they work in the agriculture farm field.

The underlying resolution of high quality is the fundamental key for providing accurate data. It is an input for the custom-made algorithm and the reports with recommendations and predictions are given to the farmers, and this entire system can be integrated into the precision agriculture machinery. They likewise give incredible guides to the producer for leading ground-truth meetings against the underlying information. By giving them, alongside sending them through their custom handling, they save cultivators the time and exertion needed to re-decipher maps for their hardware.

The algorithm developed with various options opened which helps in communicating with various farm systems. It helps in integrating with data of the weather, crop, pests, and soil which helps in expanding the ecosystem of the innovative technology. The objective is to expand this work process to cultivators which assists with extending this technology. Overcoming any barrier of this multispectral innovation to the genuine application started in the last decade and now they are perceived as one of the profoundly advancing innovations.

11.1.4 ALGORITHMIC ANALYSIS

11.1.4.1 Plant Indices

NDVI (Normalized Difference Vegetation Index) is illustrated in Figure 11.1.
NDVI (Normalized Difference Vegetation Index) is illustrated in Figure 11.2.

FIGURE 11.1   NDVI.

$$(NIR - RED)/(NIR + RED)$$

FIGURE 11.2   GNDVI.

11.1.4.2 Uses

- It helps in assessing plant health in the field.
- It helps in assessing the chlorophyll content.

GNDVI – Green NDVI is illustrated in Figure 11.3.

$$(NIR - GREEN)/(NIR + GREEN)$$

11.1.4.3 Description

GNDVI is the modified version of standard NDVI, which is more dynamic and has a more range of values than NDVI [16]. It is multiple times more delicate to chlorophyll concentration which can detect and differentiate between the healthy and unhealthy crops. It is utilized to detect the grouping of chlorophyll, to gauge the pace of photosynthesis rate of health crops with all the requirements added, and is one of the major applications in detecting the crop stress.

  CIR Composite (Color Infrared) is illustrated in Figure 11.3.

11.1.4.4 Uses

- It helps in assessing plant health in the field.
- Identifies water bodies and helps in improving draining facilities.
- Detects the variability in soil moisture and recommends where the land is dry.
- It helps in assessing soil composition by the chlorophyll content and recommend where the deficiency is low.

11.1.4.5 Description

The layer generated from the multispectral image data is a composite of color, and it is not plant vegetation index. In general, it is described as a Color Infrared Composite

FIGURE 11.3 CIR composite.

in light of the fact that as opposed to consolidating Red, Green, and Blue, NIR, Red, and Green groups are joined. NIR light is shown as red, red light is shown as green, and green light is shown as blue. This shading composite identifies the reaction of the NIR band to detect the health of crop and identify water bodies on the surface [17].

When all said is done, healthy crop data mirrors a significant level of degree of NIR and red in CIR layers. Diseased crop data will reflect less in the NIR [18] and show up as cleaned out pink tones, sick or shadow vegetation is regularly green, and human designs are light blue green. Soils show up light blue or green in color, depend upon how sandy the soil is, with the sandiest soil showing up light clay soils as dull pale blue green. This is exceptionally helpful in distinguishing water bodies in the image data, which assimilate NIR frequencies and seem dark when water is clear. As this isn't a plant index, there is no color palate range to choose for the layer of plant index. The color seen is an after effect of combination of NIR, red, and green frequencies at each image pixel.

NDRE (Normalized Difference Red Edge) is illustrated in Figure 11.4.

NDRE (Normalized Difference Red Edge) is illustrated in Figure 11.5.

$$(NIR - RE) / (NIR + RE)$$

FIGURE 11.4   NDRE.

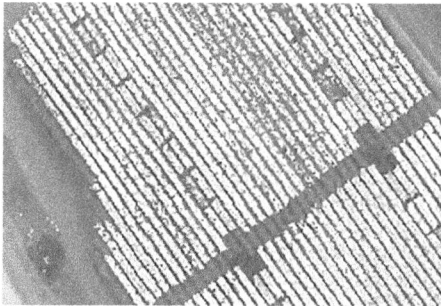

FIGURE 11.5   Chlorophyll map.

11.1.4.6   Uses

- Identifies leaf chlorophyll content.
- Detects stress and recommends suitable action.
- Recommends the deficiency of nitrogen and give the quantity that is to be added.

11.1.4.7   Description

NDRE is one of the important indices which can be developed only when the Red edge band is present in the multispectral camera. It is highly delicate to chlorophyll presence in leaves, fluctuation in leaf areas, and impacts of soils in the farm field. Higher estimations of NDRE within address more elevated levels of leaf chlorophyll content and vice versa. In general, soil demonstrates the least value qualities, and also undesirable plants or chlorophyll fading leaves have moderate range of value, and well-nourished crop with good chlorophyll content have the most elevated qualities [19–21]. Think about utilizing NDRE layer in the application management software to increase the nutrient requirement, i.e. foliar nitrogen, chlorophyll has greatest assimilation in the red waveband of the visible spectrum, and in this way red light doesn't infiltrate into leaf layers due to maximum amount of absorption. Then again, light in the green and red edge of visible spectrum can enter a leaf substantially more profoundly than blue or red light of visible spectrum so an unadulterated red-edge waveband will be more sensitive from medium to undeniable degrees of chlorophyll content, and henceforth leaf nitrogen, than a wide waveband that envelops blue light, red light, or a combination of obvious and NIR.

NDRE is a preferable marker of vegetation health over NDVI for mid to late preparation of crops that have aggregated significant degrees of chlorophyll in their leaves since red-edge light is clearer to leaves than red light, thus it is more averse to be totally consumed by an overhang. It is more reasonable for concentrated administration applications all through the developing season in light of the fact that NDVI frequently loses affectability to chlorophyll and does not foresee minute changes after plants gather a basic degree of leaf chlorophyll content [22].

11.1.4.8   Chlorophyll Map

Chlorophyll map is illustrated in Figure 11.5.

11.1.4.9   Uses

- It identifies crops with chlorosis within the field.
- It helps detecting plant stress.
- It identifies vigorous, healthy crops from unhealthy ones.
- It estimates chlorophyll content.
- It estimates nitrogen content if it is limiting within the soil.

FIGURE 11.6 OSAVI.

11.1.4.10 Description

The Chlorophyll map is a layer generated with availability of red edge sensor which is less delicate to leaf area than NDRE. It separates the chlorophyll signal from changeability in leaf region as a component of changes in overhang cover. It has a physiological premise that related the connection between chlorophyll content in farm field, identifying the nutrient presence in the crops.

It is particularly sensitive to all-round accumulated and calibrated data from multispectral camera sensor of the crop. Soil pixels and other non-vegetative pixels are removed and appeared as clear, which sometimes brings about plant pixels additionally being eliminated. This layer is less helpful for line yields and more valuable for grape plantations and farms, as the thick crop density is better at separating the chlorophyll signal [23].

OSAVI (Optimized Soil-Adjusted Vegetation Index) [24] is illustrated in Figure 11.6.

11.1.4.11 Uses

- It differentiates soil pixel to crop pixel data.
- It works at some levels where NDVI saturates when the canopy grows and chlorophyll content increases.
- It represents non-direct communications of light among soil and vegetation.
- It functions as a primary list for some joined files intended for chlorophyll location.

11.1.4.12 Explanation

This plant index in general is used in farms where there is more empty space. OSAVI maps contrast in density of the crops in the field. Moreover, it isn't delicate to soil brightness compared to other indices when distinctive soil types are available in the

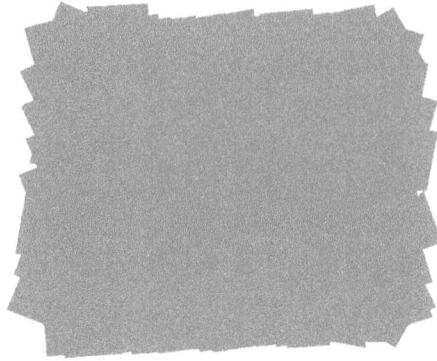

FIGURE 11.7   DSM.

farm field. It is vigorous to inconstancy in soil brightness and has upgraded its sensitivity to vegetation cover more prominent than half of the farm field. This plant vegetation index is best utilized in regions with moderately scanty vegetation where soil is noticeable through the farm field and where NDVI is unable to predict the accurate data.

OSAVI is the next version of the Soil Adjusted Vegetation Index (SAVI). It was created by utilizing the reflectance in the near-infrared (NIR) and red band of visible spectrum with an optimized soil adjustment coefficient with positive and negative values. It assists with separating the soil with the crop vegetation particularly in columns of orchids. The soil coefficient with significant range of value is 0.16 was chosen as the ideal worth to limit NDVI's affectability to variety in soil foundation under a wide scope of natural conditions. It is a mixture between proportion-based records, for example, NDVI and symmetrical files like PVI. SAVI has a default soil-change factor of 0.5; in any case, it is prescribed to utilize 0.16 as executed in OSAVI. Its esteems are in the range from −1 to 1 like NDVI and it also shows negative value when there is presence of soil compared to crop vegetation. More the positive values of OSAVI esteems show denser, better vegetation where lower esteems demonstrate less dense zone in the field [25].

Digital Surface Model (**DSM**) is illustrated in Figure 11.7.

11.1.4.13   Explanation

It can be described as computerized model presentation of the outside landscape that presents heights above sea level of the ground and highlights on it. It is a gridded cluster of heights. It is a layer represented by a gray color presentation, with enhancing the effects, for example, slope concealing might be utilized to reproduce alleviation. It assists with contemplating surface properties and draining facility flow [26].

An advanced surface model is normally built and created from automatic programmed extraction algorithms and seems covering a layer on your surface. It shows the uneven fields and all items on the landscape, including vegetation and man-made highlights, and features the various heights on a superficial level among the highlights.

To produce the recommendations from the image data, distinct, innovative algorithms are designed and developed for plant index layer, going farther than regular research and industry standardize normalized difference vegetation index (NDVI) indicators, since the information about the crop is limited. Plant lists changes from yields to crops. Measure biomass advancement, chlorophyll content and consolidates this with other field information such as soil type, crop assortment and so forth to make treatment proposals with square-meter accuracy.

11.1.4.14 Uses

- It helps to estimate relative crop volume.
- Identifies surface properties and helps in estimating the water flow.
- Models the drainage facilities in the field.

11.1.5 REAL-WORLD RESULTS

In order to give the accurate nitrogen requirement prediction, a reference fertilization model will be developed considering the local weather soil, pests, and rainfall data with advanced machine learning algorithms. The recommendations made using the reference fertilization model changes from area to area with small changes and are developed along with local research institutes. Flights over a farm depend on the type of crop in general with an interval of 7–10 days, supply the farmer with variable plant indices application layers featuring its uniquely crafted nitrogen predictions and recommendations. The nitrogen recommendation layer is a detailed version of image-processed data, developed involving the huge data of a single crop and evolved further into an automatic process by an application which creates the report in just a few minutes, which will help the farmers work with advanced precision agriculture machines and system [27]. These results are also suitable for farmers who add the nitrogen in person or utilize less advanced hardware; any grower can utilize the variable rate plant index application layer and can use the machinery depending on the recommendation in the report generated. NDRE results in the paddy field are illustrated in Figure 11.8.

FIGURE 11.8 NDRE results in the paddy field.

Fixing and deciding the ideal nitrogen (N) rate in crop fields remains a primary issue, mainly due to unaccounted spatial data, for example, properties of soil and temporal inconsistent issues like climate change. UAVs integrated with multispectral sensors provide feasible options to improve nitrogen system organization by informing of spatial variable timely; in-season N applications at the important time in the crop life cycle is helpful and plays a crucial role for the growth of crop and increase in yield. Recently, a practical decision support system (DSS) was developed to translate spatial field characteristics and normalized difference red edge (NDRE) values into an N application recommendation during the in-season. It was developed on-farm strip-trials at three different sites more than three years to differ with farmer's traditional nitrogen management with a split-application to build nitrogen management in the software guided by our UAV sensor-based DSS. The proposed framework created a system where nitrogen utilization with no yield changes compared to the farmers' regular nitrogen management system. It distinguished five avenues for additional improvement of the proposed DSS: meaning of the primary amount nitrogen rate, assessment of inputs for designing sensor algorithms, management zone outline, high-resolution image normalization standard approach, and the threshold for setting nitrogen application. Two virtual reference (VR) methods were developed where the high nitrogen reference strategies are used for normalizing high-resolution sensor image data. The VR strategies came out with a significantly lower sufficiency index range of values than those by the HN reference, resulting in N fertilization recommendations. The use of small HN reference blocks in contrasting management zones may be more appropriate to translate field-scale, high-resolution imagery into in-season N recommendations. Considering a developed interest in utilizing UAVs in business fields and the need to improve the productivity of the farm, further work is expected to refine approaches for making the valuable crop image data into in-season N recommendation reports.

11.1.5.1 Stress Management

Stress is a physiological condition restraining plants from arriving at their full genomic potential. During the crop cycle, early prediction of the stress of crops in the farm is indispensable for keeping up biomass and water retention. By using multispectral technology, growers and researchers can effectively measure relative chlorophyll content and nitrogen status and identify infection focal points to prevent stress from damaging plants. Agricultural imager analytics and drone services have been evaluating such strategies using drone-based data.

NDRE is a plant vegetation index that can be planned if red edge band sensor is available in the camera sensor. This range of bands in the visible spectrum is highly sensitive to chlorophyll content in leaves, detects the variability of chlorophyll in leaves, and also minimizes the soil effects in the background in its layer. It helps in finding the crop condition through the chlorophyll content. When NDRE shows elevated values in its metric, it represents higher levels of leaf chlorophyll content and vice versa. In general, Soil normally has the lowest values in the NDRE index, unhealthy or infected plants have intermediate values in the measuring scale, and healthy plants have the elevated values. Utilizing NDRE maps recommends the fertilizer requirements especially nitrogen variation and also predicts the availability of nitrogen content in the soil.

In the visible spectrum of light, in the red wave band, chlorophyll has maximum absorption leading to less penetration of light into deep layers of the lead of crop. Then again, light in the green and red-edge wavelength can infiltrate a leaf substantially more profoundly than blue or red light so an unadulterated red-edge waveband being sensitive to the higher levels of chlorophyll content, and hence leaf nitrogen, does not show its reflection in broad wavelengths that includes blue light, red light, or a combination of visible and NIR light.

The spatial and temporal data changes show the variation in yield assessment that are recognizable through crop biological attributes observed at various phenological developmental levels of the crop cycle. A multispectral red-edge sensor integrated into UAVs provides spatial and temporal data with high resolution which helps in providing estimations and predictions accurately. The multispectral analysis of UAVs collected spatiotemporal image data used in developing a statistical model which helps in predicting crop productivity dependent on various phenological levels. Distinguishing basic vegetation plant records and image spectral data could prompt to predict the crop yield precisely. The target of this examination was to build up a yield forecast model at explicit phenological stages utilizing image data acquired from a crop field. The accessible otherworldly groups like red, blue, green, near-infrared (NIR), and red edge were utilized to examine the excess of 30 different plant vegetation indices. The data was gathered from a field utilizing a multispectral red-edge sensor, mounted on a UAV. The research by making changes in the camera sensors, calibrating them and others to reduce the issues while gathering the image data. This outcome in the new observational technique used to decrease the impact pixels of bare soil in image data obtained with the exploratory design.

11.1.5.2 Nitrogen Recommendation Methodology

The methodology of treatments of nitrogen (N) with different amounts was applied in a random manner. The random field was used as a feature selection method to choose the combination of different parameters for various stages of the crop. Multiple linear regression of the image data was used to develop yield prediction models for each specific phenological stage by utilizing the most effective parameter at each stage of the crop life cycle. At the various stages of the crop cycle, the Optimized Soil Adjusted Vegetation Index (OSAVI) and Simplified Canopy Chlorophyll Content Index (SCCCI) were the single prevailing factors in the yield predicting models of the system, individually. The combined layers of the Normalized Difference Red-Edge, and green Normalized Difference Vegetation Index (GNDVI) at the developing stages, and Optimized Soil-Adjusted Vegetation Index (OSAVI) at the stage before the measure of nitrogen to be applied, and SCCCI, Green Leaf Index (GLI), Optimized Soil Adjusted Vegetation Index (OSAVI) and Visible Atmospherically Resistant Index (VARI green) at tasselling stage (VT) were the best plant vegetation indices for predicting grain yield of the crop. In addition, the SCCCI as a combined plant vegetation index seemed to be the important record of plant parameters for anticipating yield considering the phenological stages of crop.

Using the multispectral camera to monitor and detect plant physiological conditions, the team set out to pinpoint the first signs of stress within a wheat crop. Researchers measured relative chlorophyll content for plant stress, identified areas low in nutrients, and detected infection focal points. The data collected was analyzed using their online platform.

11.1.5.3 Analyzing the Effects of Fertilizers in Plant Chlorophyll Levels

Analyzing the reaction of crops to different levels of fertilizers throughout the vegetative season reflects the changes in the chlorophyll content. The multispectral camera helps in easy identification of the area in the farm with low availability of nutrients and monitors the crop and recommends the fertilizer applications throughout the stages of the crop cycle and helps in recommending where new fertilizers are necessary at the given chunk of the farm.

With the Soil Plant Analysis Development Chlorophyll Meter and the camera sensor, Treatments were adjusted based on intensive fertilizing methods available for enabling calibration of chlorophyll measurements from long-range spectroscopy-based plant physiology research. As a result, it was possible to separate and evaluate nutrient-deficient plots. Some plots showed earlier signs of deficiency, so a species-specific aspect of the nutrient reaction was also studied.

To visualize differences between plots, False Coloring methods are used. The color differences are basic indicators of crop health status. For example, differing colors inside the crop columns below show discrepancies among chlorophyll levels, with red spots indicating chlorophyll deficiency. With such data, it is easy to develop new indices, such as the Chlorophyll Index.

11.1.5.4 Detection of Disease Using the Red Edge Band

Likewise, with most of the infections affecting productivity, early detection of disease is urgent. To accomplish that, the multispectral sensor is crucial, yet the infection to the crop stayed undetectable until featured by the Chlorophyll map which is a red edge-based plant vegetation index. The crop stress may occur due to water inadequacy, nitrogen deficiency, and fungal infection, early detection, and quick reaction are vital to forestall the infection which directly affects the yield and gain the potential yield maximum. A sensor without the red edge band can detect the stress and disease in the farm fields, but it may detect when the disease spread in the field was using the Chlorophyll map it may help to recover the potential crop yield as a result in an increase in profit per given section of land.

The research involves increasing potential yield, productivity, gaining profit, and agriculture with sustainable technology by using robots, multispectral analysis, and others. By collecting the image data from the field, Researchers review the multispectral image data, and it is identified that a multispectral camera with red-edge wavelength can recognize the stress of crop before the RGB or an NDVI-based camera sensor since the Red-Edge camera edge of the wavelength of the red color of the light spectrum which is a critical region in detecting various issues of the crop especially in dealing with the chlorophyll), and it is known as the red edge band (712–722) nm. This is the important band in the visible spectrum which identifies the crop stress.

Utilizing the plant vegetation indices at the point like NDRE or Chlorophyll map, the red edge band helps in identifying disease earlier depending on the chlorophyll content. Generally, if the disease is pervasive in a yield, it tends to be found in red edge-based plant vegetation indices before it gets noticeable in NDVI. Early detection assists farmers to catch disease sooner and to take appropriate action without spreading the entire field. It helps act quicker to stop the spread.

The Red-Edge camera sensor is used to identify disease in the crops before it was visible in NDVI. The data of plant indices designed from Red-Green-Blue (RGB), normalized difference vegetation index, and CIR Composite layers where there is no RED edge in the formula, the farm field appeared healthy and unvaried. But it helps to finds in establishing the reason for causing crop stress in the field of the particular region. When we use red edge sensor data, it shows the stress in patches in the field and going through the area showing the disease symptoms and can be explained due to the decrease in the chlorophyll content in the crop, and it can also be seen in the Chlorophyll map where the red-edge sensor is used. From the analysis, we can see the area where it got infected and taking the action within the field. These applications facilitate farmers in reducing the chances of infections that impact the farm produce.

11.1.5.5 Analyzing the Effects of Various Fungicides on Disease Management

The researchers analyzed the response of various types of crops to fungicides, focusing on resistance against pathogens in general. Some of these groups were treated with fungicide and others were not. The lack of fungicide treatment in some groups led to severe infections. Infected plants display symptoms like white powdery spots. Using the multispectral camera, the team could identify infection focal points in the field and provide information about which part of the field contained susceptible species. With the data captured by the multispectral camera, it is able to identify and monitor nutrient deficiency and to detect, treat, and analyze the response to pathogen infection. The success of such experiments proved the multispectral camera an effective tool for monitoring the physiological conditions of crops.

Generally, controlling or identifying the disease in the field is not possible by walking in the fields, due to the symptoms of disease seen by the human naked eye in the fields are not detected. The image data collected from the UAVs integrated with multispectral cameras help in developing the recommended layers of plant vegetation indices to detect the leaf diseases in the crop in the entire field. It recommends to spray the pesticides in precise quantity to the entire field. The prediction develops by taking the set of images infected with the disease and another set of healthy green leaf in establishing the threshold values. The reflectance values of selected images are used to develop 18 vegetation indices, and these indices are used to calculate the percentage difference of vegetation by comparing with the healthy leaf, and plant indices reports are generated to the unhealthy leaf along with recommendation to the value how much amount of pesticides are sprayed per hectare. The plant indices result in general has NIR and red edge band in its formula most of its plant indices that have a different percentage in the range of 15–45%, with NDRE, Green NDVI showing the huge percentage difference and these plant indices using only visible bands in the formula of four vegetation indices have the percentage difference in the range of 25% with GI and NRI results with the highest difference percentage around 26%.

11.1.5.6 Detection of Weeds

The detection of weeds, in the present times, is a distinct challenge to farmers. Buried in the field and growing in the same way as the remainder of the field, the weeds can be difficult to spot. Once established, weeds can have a negative impact on farming operations since like any other weed, are competing with planted crops for water, nutrients, light, and space. Moreover, the weeds produce toxins that inhibit the crop growth, but that can end up being low-grade output from the farm field, causing customers and distributors to become weary of purchasing from that grower.

For one ranch in the presence of weeds has been an item of concern since it exists with crops that sometimes look identical to and mature at a similar time as the planted variety. The ability to differentiate and identify where these weeds exist has become a priority for the grower and using layers within image processing software has become a key strategy in spotting and removing the weeds. Looking closely at the weeds, it is noticeable with the size, shape of leaves that compete with the crops. In contrast, the planted variety has different size, shape with different texture, an appearance, and bark. These differences are hard to see at glance, but the greenness variation between the two types of leaves could be something easy to identify using multispectral data. The mapping of the field using the multispectral camera and analyzing the imagery collected are processed using the software.

Using a field where weeds had already been spotted on the ground, the first step was to identify, mark, and GPS tag the weeds in the multispectral layer. Then, a different methodology should be applied when the weeds are not identified in the field. The weed detection layers will be developed and generated for the farmers and agronomists to detect and varying the weeds from crops grown based on chlorophyll content in the plants. Each layer has a different combination of bands and color composite that not only make it possible to differentiate between plant types but can also help in early chlorosis detection which is difficult to see with the naked eye.

Using the chlorophyll detection layer, which identifies variation in chlorophyll content, it is easy to identify all the marked weeds and even spotted an unknown weed in the field. With a method in place, the weed detection layers used not only proved effective in identifying the weeds but also resulted in reduced cost. Having a scout walking one of the smaller fields and manually identifying the weeds would have cost a total heavily along with the time consumption. Instead, using a multispectral camera and analysis can save time, money and prevents a significant yield loss. Identifying and controlling the weeds would have been detrimental to the farm, not only for the impact on yield but the potential loss for the farmers for the upcoming seasons. The use of multispectral analysis can help farmers make more informed decisions, better allocate their resources, be more efficient in farming operations, and implement the right strategies that would result in yield and quality improvement.

11.2 RESULTS

Utilizing the drone integrated with the multispectral camera has many advantages especially since fertilizer application maps have had a good effect on farmer's operations. From the analysis, the growers will get a median output increase of 10–15% compared to output generated analyzed using ancient, non-drone techniques and strategies. This increase in productivity in the field is of huge value. The results generated are approximately for every producer in the field, and they expect to follow the practices of machine learning methods. Considering the production yield depending on many variables: from the weather, soil types, pests in particular areas, and a plant's particular trait results are a solid, real-world example of how drone data and expert algorithmic analysis can have a real beneficial effect on farmers' businesses. Today multispectral analysis of the agricultural map, serving farmers in taking the decision, researchers to accumulate statistics of data and helping them to develop advanced algorithms.

11.3   CONCLUSION

Multispectral analysis plays an evolutionary role in developing precision agriculture methods by reducing water usage, detecting pests before the damage occurs, and precise nitrogen recommendations. It may enhance the productivity of the farm. It even plays an important role in predicting the crop harvesting time from the data. Multispectral analysis is presently increasing the productivity and reducing the costs for the farmers holding large agriculture farm fields. Recent research has increased in recent times to use fixed-wing drones and advanced GPS techniques like RTK (Real-Time Kinematics) for small-scale farm holders to reduce the cost of investment and also increase the land covered in a single flight [26], which has reduced by simulating the requirement in software and meeting requirements in a manner by overcoming the drawbacks. Principles like AI and machine learning can be implemented to enhance the accuracy of recommendations for farmers.

REFERENCES

[1] Albetis, J., Jacquin, A., Goulard, M., Poilvé, H., Rousseau, J., Clenet, H., … Duthoit, S. (2018). On the potentiality of UAV multispectral imagery to detect flavescence dorée and grapevine trunk diseases. *Remote Sensing*, 11(1), 23. Retrieved March 30, 2021, from http://dblp.uni-trier.de/db/journals/remotesensing/remotesensing11.html

[2] Chen, Z., & Wang, X. (2019). Model for estimation of total nitrogen content in sandalwood leaves based on nonlinear mixed effects and dummy variables using multispectral images. *Chemometrics and Intelligent Laboratory Systems*, 195, 103874. Retrieved March 30, 2021, from https://sciencedirect.com/science/article/pii/s016974391930485x

[3] Assmann, J. J., Kerby, J. T., Cunliffe, A. M., & Myers-Smith, I. H. (2018). Vegetation monitoring using multispectral sensors – best practices and lessons learned from high latitudes. *bioRxiv*, 334730. Retrieved March 30, 2021, from https://biorxiv.org/content/biorxiv/early/2018/05/30/334730.full.pdf

[4] Baerdemaeker, J. D., & Baerdemaeker, J. D. (2013). *Multiscale photonics for precision agriculture*. Retrieved March 30, 2021, from https://spiedigitallibrary.org/conference-proceedings-of-spie/8881/1/multiscale-photonics-for-precision-agriculture/10.1117/12.2032136.full

[5] Cao, S., Danielson, B., Clare, S., Koenig, S., Campos-Vargas, C., & Sanchez-Azofeifa, A. (2019). Radiometric calibration assessments for UAS-borne multispectral cameras: Laboratory and field protocols. *Isprs Journal of Photogrammetry and Remote Sensing*, 149, 132–145. Retrieved March 30, 2021, from https://sciencedirect.com/science/article/pii/s0924271619300267

[6] Gonzalez, A., Moreno, J., Russell, G., & Marquez, A. (2009). *Using Kernel Methods in a Learning Machine Approach for Multispectral Data Classification. An Application in Agriculture.* Retrieved March 30, 2021, from https://intechopen.com/books/geoscience-and-remote-sensing/using-kernel-methods-in-a-learning-machine-approach-for-multi-spectral-data-classification-an-applica

[7] Harris, R. (2003). Remote sensing of agriculture change in Oman. *International Journal of Remote Sensing,* 24(23), 4835–4852. Retrieved March 30, 2021, from https://tandfon-line.com/doi/full/10.1080/0143116031000068178

[8] Hollaus, F., Gau, M., & Sablatnig, R. (2013). *Enhancement of Multispectral Images of Degraded Documents by Employing Spatial Information.* Retrieved March 30, 2021, from http://dblp.uni-trier.de/db/conf/icdar/icdar2013.html

[9] Huang, Y., Lan, Y., & Hoffmann, W. C. (2008). Use of airborne multi-spectral imagery in pest management systems. *Agricultural Engineering International: The CIGR Journal,* 10. Retrieved March 30, 2021, from http://cigrjournal.org/index.php/ejounral/article/view/1019

[10] Iannini, L., Molijn, R. A., & Hanssen, R. F. (2013). *Integration of multispectral and C-band SAR data for crop classification.* Retrieved March 30, 2021, from https://narcis.nl/publication/recordid/oai:tudelft.nl:uuid:14e44d11-4561-4211-b302-3a3cd3b33b26

[11] Khelifi, R., Adel, M., & Bourennane, S. (2012). Multispectral texture characterization: application to computer aided diagnosis on prostatic tissue images. *EURASIP Journal on Advances in Signal Processing,* 2012(1), 118. Retrieved March 30, 2021, from https://link.springer.com/content/pdf/10.1186/1687-6180-2012-118.pdf

[12] Lussem, U., Bolten, A., Menne, J., Gnyp, M. L., & Bareth, G. (2019). Ultra-high spatial resolution UAV-based imagery to predict biomass in temperate grasslands. *ISPRS - International Archives of the Photogrammetry, Remote Sensing and Spatial Information Sciences,* 443–447. Retrieved March 30, 2021, from https://int-arch-photogramm-remote-sens-spatial-inf-sci.net/xlii-2-w13/443/2019/isprs-archives-xlii-2-w13-443-2019.pdf

[13] Marinello, F. (2017). Last generation instrument for agriculture multispectral data collection. *Agricultural Engineering International: The CIGR Journal,* 19(1), 87–93. Retrieved March 30, 2021, from https://cigrjournal.org/index.php/ejounral/article/view/3939/2517

[14] Moody, D. I., Brumby, S. P., Chartrand, R., Keisler, R., Longbotham, N., Mertes, C., … Warren, M. S. (2017). Crop classification using temporal stacks of multispectral satellite imagery. *Proceedings of SPIE,* 10198. Retrieved March 30, 2021, from https://spiedigi-tallibrary.org/conference-proceedings-of-spie/10198/1/crop-classification-using-tem-poral-stacks-of-multispectral-satellite-imagery/10.1117/12.2262804.full

[15] Pop, N., & Toderas, T. (2014). *Some Considerations Regarding the Usage of Multispectral Remote Sensing Images in Agricultural Crop Analysis.* Retrieved March 30, 2021, from https://dissem.in/p/84420969/some-considerations-regarding-the-usage-of-multispectral-remote-sensing-images-in-agricultural-crop-analysis

[16] Neely, H. L., Morgan, C. L., Stanislav, S. M., Rouze, G., Shi, Y., Thomasson, J. A., … Olsenholler, J. (2016). *Strategies for soil-based precision agriculture in cotton.* Retrieved March 30, 2021, from https://spiedigitallibrary.org/conference-proceedings-of-spie/9866/1/strategies-for-soil-based-precision-agriculture-in-cotton/10.1117/12.2228732.short

[17] Ramos, S. Z. (2017). *Comparison of multi-temporal and multispectral Sentinel-2 and Unmanned Aerial Vehicle imagery for crop type mapping.* Retrieved March 30, 2021, from http://lup.lub.lu.se/student-papers/record/8917610/file/8917627.pdf

[18] Ren, D. D., Tripathi, S., & Li, L. K. (2017). Low-cost multispectral imaging for remote sensing of lettuce health. *Journal of Applied Remote Sensing*, 11(1), 016006–016006. Retrieved March 30, 2021, from https://spiedigitallibrary.org/journals/journal-of-applied-remote-sensing/volume-11/issue-01/016006/low-cost-multispectral-imaging-for-remote-sensing-of-lettuce-health/10.1117/1.jrs.11.016006.full

[19] Singh, P., Gupta, A., & Singh, M. (2014). Hydrological inferences from watershed analysis for water resource management using remote sensing and GIS techniques. *The Egyptian Journal of Remote Sensing and Space Science*, 17(2), 111–121. Retrieved March 30, 2021, from https://sciencedirect.com/science/article/pii/s1110982314000271

[20] Sofonia, J., Shendryk, Y., Phinn, S. R., Roelfsema, C., Kendoul, F., & Skocaj, D. (2019). Monitoring sugarcane growth response to varying nitrogen application rates: A comparison of UAV SLAM LiDAR and photogrammetry. *International Journal of Applied Earth Observation and Geoinformation*, 82, 101878. Retrieved March 30, 2021, from https://sciencedirect.com/science/article/pii/s0303243418312522

[21] Solecki, C. F. (2017). *Evaluating Unmanned Aerial Vehicle Based Crop Indexing Techniques: Modified Consumer Grade RGB Vs. Multispectral.* Retrieved March 30, 2021, from https://tru.arcabc.ca/islandora/object/tru:1338/datastream/pdf/view

[22] Sosa-Herrera, J. A., Vallejo-Pérez, M. R., Alvarez-Jarquin, N., Cid-García, N. M., & López-Araujo, D. J. (2019). Geographic object-based analysis of airborne multispectral images for health assessment of capsicum annuum L. crops. *Sensors*, 19(21), 4817. Retrieved March 30, 2021, from https://mdpi.com/1424-8220/19/21/4817

[23] Stroppiana, D., Pepe, M., Boschetti, M., Crema, A., Candiani, G., Giordan, D., ... Monopoli, L. (2019). Estimating crop density from multi-spectral uav imagery in maize crop. *ISPRS - International Archives of the Photogrammetry, Remote Sensing and Spatial Information Sciences*, 619–624. Retrieved March 30, 2021, from https://int-arch-photogramm-remote-sens-spatial-inf-sci.net/xlii-2-w13/619/2019/isprs-archives-xlii-2-w13-619-2019.pdf

[24] Su, J., Liu, C., Coombes, M., Hu, X., Wang, C., Xu, X., ... Chen, W.-H. (2018). Wheat yellow rust monitoring by learning from multispectral UAV aerial imagery. *Computers and Electronics in Agriculture*, 155, 157–166. Retrieved March 30, 2021, from https://sciencedirect.com/science/article/pii/s0168169918312584

[25] Thomson, S. J., & Sullivan, D. G. (2006). *Crop Status Monitoring using Multispectral and Thermal Imaging Systems for Accessible Aerial Platforms.* Retrieved March 30, 2021, from https://elibrary.asabe.org/abstract.asp?aid=20657

[26] Thwal, N. S., Ishikawa, T., & Watanabe, H. (2019). Land cover classification and change detection analysis of multispectral satellite images using machine learning. Retrieved March 30, 2021, from https://spiedigitallibrary.org/conference-proceedings-of-spie/11155/2532988/land-cover-classification-and-change-detection-analysis-of-multispectral-satellite/10.1117/12.2532988.short

[27] Zhang, H., Lan, Y., Lacey, R. E., Huang, Y., Hoffmann, W. C., Martin, D., & Bora, G. C. (2009). Analysis of variograms with various sample sizes from a multispectral image. *International Journal of Agricultural and Biological Engineering*, 2(4), 62–69. Retrieved March 30, 2021, from https://ijabe.org/index.php/ijabe/article/viewfile/201/98

Index

For Product Safety Concerns and Information please contact our EU
representative GPSR@taylorandfrancis.com
Taylor & Francis Verlag GmbH, Kaufingerstraße 24, 80331 München, Germany